T0183066

SpringerBriefs in Electrical and Computer Engineering

More information about this series at http://www.springer.com/series/10059

Yufei Jiang • Xu Zhu • Eng Gee Lim
Yi Huang • Hai Lin

Semi-Blind Carrier Frequency Offset Estimation and Channel Equalization

 Springer

Yufei Jiang
University of Liverpool
Liverpool, Merseyside, UK

Xu Zhu
University of Liverpool
Liverpool, Merseyside, UK

Eng Gee Lim
Xi'an Jiaotong-Liverpool University
Suzhou, China

Yi Huang
University of Liverpool
Liverpool, Merseyside, UK

Hai Lin
Osaka Prefecture University
Osaka, Japan

ISSN 2191-8112   ISSN 2191-8120   (electronic)
SpringerBriefs in Electrical and Computer Engineering
ISBN 978-3-319-24982-7   ISBN 978-3-319-24984-1   (eBook)
DOI 10.1007/978-3-319-24984-1

Library of Congress Control Number: 2015953666

Springer Cham Heidelberg New York Dordrecht London
© The Author(s) 2015
This work is subject to copyright. All rights are reserved by the Publisher, whether the whole or part of the material is concerned, specifically the rights of translation, reprinting, reuse of illustrations, recitation, broadcasting, reproduction on microfilms or in any other physical way, and transmission or information storage and retrieval, electronic adaptation, computer software, or by similar or dissimilar methodology now known or hereafter developed.
The use of general descriptive names, registered names, trademarks, service marks, etc. in this publication does not imply, even in the absence of a specific statement, that such names are exempt from the relevant protective laws and regulations and therefore free for general use.
The publisher, the authors and the editors are safe to assume that the advice and information in this book are believed to be true and accurate at the date of publication. Neither the publisher nor the authors or the editors give a warranty, express or implied, with respect to the material contained herein or for any errors or omissions that may have been made.

Printed on acid-free paper

Springer International Publishing AG Switzerland is part of Springer Science+Business Media (www.springer.com)

Preface

The technologies of wireless communications have experienced a rapid growth over the past two decades. The demands for high-data-rate services have motivated numerous research activities to be carried out. However, the demands are limited by the very scarce bandwidth resource. Commonly, Channel State Information (CSI) and Carrier Frequency Offset (CFO) estimations are performed by using training signals, which reduce the spectral efficiency further. Therefore, it is very urgent and important to improve the bandwidth usage.

Independent Component Analysis (ICA) is an efficient Higher-Order Statistics (HOS)-based blind source separation technique by maximizing non-Gaussianity of the ICA output signals. So far, ICA has been applied to a range of fields, including separation of signals in audio applications or brain imaging, the analysis of economic data, and feature extraction. Since the use of ICA has the benefit of not requiring the Channel State Information (CSI) to perform blind or semi-blind equalization, it has been proven to be effective for a number of wireless communications systems, like blind channel estimation, blind equalization, and blind multiuser detection. Thus, it is very interesting to investigate the application of ICA to a number of OFDM-based wireless communication systems.

In this book, we apply ICA for a number of OFDM-based wireless communication systems, with the effect of CFO. In Chap. 1, a few concepts are introduced, such as Orthogonal Frequency Division Multiplexing (OFDM), equalization, CFO, etc. In Chap. 2, a number of wireless communication systems are introduced, such as Multiple-Input Multiple-Output (MIMO), Coordinated Multi-Point (CoMP), and Carrier Aggregation (CA). In Chap. 3, some existing preamble-based and blind CFO estimation methods are described. Then, a number of the training-based and blind channel estimation and equalization methods are reviewed. This is followed by a review of the basics of ICA and the application to a number of semi-blind OFDM-based wireless communication systems. In Chap. 4, a precoding-based CFO estimation method and an ICA-based equalization structure are proposed for semi-blind single-user MIMO OFDM systems, where a number of reference data sequences are superimposed on the source data sequences via a precoding process, for CFO estimation and ambiguity elimination in the ICA equalized signals. In

Chap. 5, a semi-blind multiuser CoMP OFDM system is proposed, with a low-complexity multi-CFO estimation method and an ICA-based equalization structure, where a small number of pilots are designed to perform multi-CFO estimation and ambiguity elimination in the ICA equalized signals. In Chap. 6, a semi-blind ICA-based joint ICI mitigation and equalization scheme is proposed for CA-based CoMP OFDMA systems, where the CFO-induced ICI is mitigated implicitly via an ICA-based semi-blind equalization. Finally, the findings are summarized and conclusions are drawn in Chap. 7.

Liverpool, UK   Yufei Jiang
Liverpool, UK   Xu Zhu
Suzhou, China   Eng Gee Lim
Liverpool, UK   Yi Huang
Osaka, Japan   Hai Lin

Contents

Acronyms

5G	Fifth Generation
AWGN	Additive White Gaussian Noise
BER	Bit Error Rate
BS	Base Station
BSS	Blind Source Separation
CA	Carrier Aggregation
CAI	Co-Antenna Interference
CAS	Carrier Assignment Scheme
CAZAC	Constant Amplitude Zero Auto-Correlation
CCI	Co-Channel Interference
CFO	Carrier Frequency Offset
CIR	Channel Impulse Response
CMA	Constant Modulus Algorithm
CoMP	Coordinated Multi-Point
CP	Cyclic Prefix
CSI	Channel State Information
DCA	Direct Conversion Architecture
DFT	Discrete Fourier Transform
EVD	Eigenvalue Decomposition
FA	Finite Alphabet
FDE	Frequency Domain Equalization
FDM	Frequency Division Multiplexing
FFT	Fast Fourier Transform
HOS	Higher Order Statistics
i.i.d.	Independent Identically Distributed
I/Q	Inphase/Quadrature
IBI	Inter-Block Interference
ICA	Independent Component Analysis
ICI	Inter-Carrier Interference
IDFT	Inverse Discrete Fourier Transform
IFFT	Inverse Fast Fourier Transform

ISI	Inter-Symbol Interference
IUI	Inter-User Interference
JADE	Joint Approximate Diagonalization of Eigenmatrices
LOS	Line-of-Sight
LOs	Local Oscillators
LS	Least-Square
LTE	Long Term Evolution
MIMO	Multiple-Input Multiple-Output
ML	Maximum Likelihood
MMSE	Minimum Mean Square Error
MPP	Modified Periodic Pilot
MSE	Mean Square Error
MUI	Multiple-User Interference
MUSIC	Multiple Signal Classification
NLOS	Non Line-of-Sight
OFDM	Orthogonal Frequency Division Multiplexing
OFDMA	Orthogonal Frequency Division Multiple Access
OP	Orthogonal Pilot
PAPR	Peak-to-Average Power Ratio
PCA	Principal Component Analysis
PDF	Probability Density Function
PDP	Power Delay Profile
RMS	Root Mean Square
SIC	Successive Interference Cancellation
SIMO	Single-Input Multiple-Output
SISO	Single-Input Single-Output
SNR	Signal-to-Noise Ratio
SOP	Semi-Orthogonal Pilot
SOS	Second Order Statistics
STO	Symbol Time Offset
V-BLAST	Vertical Bell Laboratories Layered Space-Time
ZF	Zero Forcing
ZP	Zero Padding

Chapter 1
Introduction

The technologies of wireless communications have experienced a rapid growth over the past two decades [1]. The demands for high-data-rate services has motivated numerous research activities to be carried out to promote a higher system capacity. To achieve the purpose, a number of wireless communication systems have been proposed and used. Multiple-input Multiple-Input Multiple-Output (MIMO) systems [2], which can employ multiple transmit and receive antennas, offer considerable capacity improvement over Single-Input Single-Output (SISO) systems. Coordinated Multi-Point (CoMP) transmission [3] can effectively manage the interference between cells and improve cell-edge throughput, by allowing each separated Base Station (BS) to jointly deal with multiple users' signals. Carrier Aggregation (CA) transmission [4, 5] can provide very high data rates, by aggregating multiple component carriers for the concurrent transmission. So far, the above techniques, have been chosen for the wireless local area networks IEEE 802.11 standards [6], and adopted by Long Term Evolution (LTE)-Advanced standards [7, 8].

Orthogonal Frequency Division Multiplexing (OFDM) technology, first proposed in 1960 [9], is robust against frequency selective fading, by dividing frequency selective fading channels into a number of frequency flat fading channels. Due to the rapid development of digital signal processing, this technology becomes practically possible and attractive. Orthogonal Frequency Division Multiple Access (OFDMA) [6], inherited from OFDM, can improve frequency diversity for users to transmit signals simultaneously, by using different subcarriers. So far, OFDM or OFDMA has been employed in a range of wireless communication systems, such as MIMO OFDM systems, CoMP OFDM systems and CA based CoMP OFDMA systems.

At the receiver, the reverse process is referred to as equalization [1], transforming the received signals back into the transmitted signals. To perform equalization, equalizer coefficients are required. Usually, they can be obtained directly from

© The Author(s) 2015
Y. Jiang et al., *Semi-Blind Carrier Frequency Offset Estimation and Channel Equalization*, SpringerBriefs in Electrical and Computer Engineering, DOI 10.1007/978-3-319-24984-1_1

Channel State Information (CSI). Traditionally, training signals are commonly used in wireless communication systems to estimate the CSI at the receiver. However, transmitting training signals reduces spectral efficiency, especially in wireless communication systems which have very scarce bandwidth resource. Blind or semi-blind channel estimation and equalization methods [10] can obtain the CSI and recover the source data directly from the structure and statistics of the received signals, without extra bandwidth and power needed for training. As no or little prior information is available in advance at the receiver, blind or semi-blind equalization approaches can improve spectral efficiency. The aim of this work is to research a number of semi-blind wireless communication systems over frequency selective channels.

Independent Component Analysis (ICA) [11], as an efficient Higher Order Statistics (HOS) based Blind Source Separation (BSS) technique, can recover the source signals by maximizing the non-Gaussianity of the observed signals. Compared to the Second Order Statistics (SOS) based blind methods, ICA potentially reduces noise sensitivity, since the fourth or higher order cumulants of the Gaussian noise are equal to zero. The aim of ICA is to find statistically independent components from the received data. So far, ICA has been applied to a range of fields, including separation of signals in audio applications or brain imaging, the analysis of economic data and feature extraction [11]. The use of ICA has the benefit of not requiring the CSI to perform blind or semi-blind equalization. To improve spectral efficiency, ICA is applied to equalize the received signals for a number of OFDM based wireless communication systems.

However, OFDM based systems have several drawbacks. One of drawbacks is the Carrier Frequency Offset (CFO) [12], caused by the unavoidable difference between Local Oscillators (LOs) at the transmitter and receiver. The CFO destroys the orthogonality between OFDM subcarriers, and results in the Inter-Carrier Interference (ICI). This effect incurs a significant degradation in Bit Error Rate (BER) performance, if not estimated and compensated for correctly. In single-user MIMO OFDM systems, only a single CFO exists between transmitter and receiver, while there are multiple CFOs in multiuser CoMP OFDM systems. The application of CA allows multiple LOs to be installed for the concurrent transmission. Thus, there are a large number of CFOs in CA based CoMP OFDMA systems, resulting in challenging multi-CFO estimation problems.

References

1. D. Tse and P. Viswanath. *Fundamentals of Wireless Communications*. Cambridge University Press, 2005.
2. G. J. Foschini. Layered space-time architecture for wireless communication in a fading environment when using multi-element antennas. *Bell Labs Technical Journal*, 1:41–59, Oct. 1996.
3. 3gpp technical report 36.814 version 9.0.0, further advancements for e-utra physical layer aspects, Mar. 2010.

4. Z. Shen, A. Papasakellariou, J. Montojo, D. Gerstenberger, and F. Xu. Overview of 3GPP LTE-Advanced carrier aggregation for 4G wireless communications. *IEEE Communications Magazine*, 50(2):122–130, Feb. 2012.
5. K. I. Pedersen, F. Frederiksen, C. Rosa, L. G. U. Garcia H. Nguyen, and Y. Wang. Carrier aggregation for LTE Advanced: functionality and performance aspects. *IEEE Communications Magazine*, 49(6):89–95, Jun. 2011.
6. Ieee standard for information technology-telecommunications and information exchange between systems-local and metropolitan area networks-specific requirements part 11: Wireless lan medium access control (mac) and physical layer (phy) specifications, Mar. 2012.
7. K. Fazel and S. Kaiser. *Multi-carrier and spread spectrum systems: from OFDM and MC-CDMA to LTE and WiMAX*. John Wiley & Sons, New York, USA, second edition, 2008.
8. A. Technologies. *LTE and the Evolution to 4G Wireless: Design and Measurement Challenges*. John Wiley & Sons, New York, USA, 2009.
9. S. Weinstein and P. Ebert. Data transmission by frequency-division multiplexing using the discrete fourier transform. *IEEE Transactions on Communications*, 19(5):628–634, Oct. 1971.
10. L. Tong and S. Perreau. Multichannel blind identification: From subspace to maximum likelihood methods. *Proceedings of the IEEE*, 86(10):1951–1968, Oct. 1998.
11. J. Karhunen A. Hyvarinen and E. Oja. *Independent Component Analysis*. John Wiley & Sons, New York, USA, May 2002.
12. Y. S. Cho, J. Kim, W. Y. Yang, and C. G. Kang. *MIMO-OFDM Wireless Communications with MATLAB*. Wiley, Singapore, 2010.

Chapter 2
OFDM Based Wireless Communications Systems

2.1 OFDM Technology

To eliminate the Inter-Symbol Interference (ISI) due to multipath, traditional time-domain equalization schemes [1 3] in the literature require complex implementation, and thus are practically impossible. Due to technological development, the digital implementation of the pair of Fast Fourier Transform (FFT) and Inverse Fast Fourier Transform (IFFT) becomes simple and practically possible. OFDM technology, proposed in 1960 [4], is attractive for simplifying the equalization process at the receiver for frequency selective fading channels [5]. So far, OFDM has been chosen for the wireless local area networks IEEE 802.11 standards [6], and has been adopted by LTE-Advanced [7, 8]. Also, it has been selected as a strong candidate for the WiGig [9, 10] and the Fifth Generation (5G) wireless communication [11, 12] in the future, respectively.

2.1.1 Principle of OFDM

Different from the single carrier modulation which has a relatively low data rate, the multi-carrier modulation [5] is employed to support high data rates in OFDM technology. In OFDM, a frequency band is divided into a number of closely spaced spectrums. Thanks to the orthogonal property of subcarriers, the center frequency of one subcarrier can coincide with the spectral zeros of all other subcarriers, leading to no interference between subcarriers. To guarantee the orthogonality between subcarriers, a small spacing between subcarriers is used in OFDM systems, while a bigger spectral width between parallel channels is used to avoid the interference

© The Author(s) 2015
Y. Jiang et al., *Semi-Blind Carrier Frequency Offset Estimation and Channel Equalization*, SpringerBriefs in Electrical and Computer Engineering,
DOI 10.1007/978-3-319-24984-1_2

in traditional Frequency Division Multiplexing (FDM) systems [13, 14]. Therefore, OFDM systems can achieve higher spectral efficiency than FDM systems.

In order to combat frequency selective fading, a high-rate transmit stream is divided into a large number of low-rate substreams. Consequently, the symbol period on each subcarrier is prolonged, and the signal bandwidth becomes shorter in comparison to the channel coherence bandwidth. Assuming that the channel variation is slow, the symbol duration T_s can satisfy

$$\sigma_{\text{rms}} < NT_s < T_c, \tag{2.1}$$

where N is the number of subcarriers, σ_{rms} and T_c are the Root Mean Square (RMS) delay spread of the channel and the channel coherent time, respectively.

Therefore, the frequency selective fading channel can be divided into a number of frequency flat fading channels. As the ISI is avoided, OFDM systems require equalization with lower complexity than that employed in traditional methods as in [1–3].

2.1.2   CFO

Although OFDM can combat frequency selective fading, there are some drawbacks to OFDM systems. One of drawbacks is the CFO. Generally, the signal is converted up to a passband by a carrier at the transmitter, and converted down to the baseband by the same carrier at the receiver, via LOs. This type of CFO is caused by the unavoidable difference of LOs between transmitter and receiver. The CFO destroys the orthogonality between OFDM subcarriers, and results in the ICI. As a result, the CFO can incur a significant degradation in BER performance.

f_r and f_t are defined as the carrier frequencies at the receiver and transmitter, respectively. The normalized CFO ϕ can be written as

$$\phi = \frac{f_r - f_t}{\Delta f}, \tag{2.2}$$

where Δf denotes the subcarrier spacing.

Define $\mathbf{s} = [s(0), s(1), \ldots, s(N-1)]^T$ as the transmitted signal vector. Considering the effect of CFO, the received signal vector $\mathbf{y}^{(\phi)} = [y^{(\phi)}(0), y^{(\phi)}(1), \ldots, y^{(\phi)}(N-1)]^T$ in the frequency domain for OFDM systems can be written as

$$\mathbf{y}^{(\phi)} = \mathbf{F}\boldsymbol{\Phi}^{(\phi)}\mathbf{F}^H\mathbf{H}\mathbf{s} + \mathbf{z}_f, \tag{2.3}$$

where $\mathbf{F}$ is an $N \times N$ Discrete Fourier Transform (DFT) matrix, with entry (a, b) given by $\mathbf{F}(a, b) = \frac{1}{\sqrt{N}}e^{\frac{-j2\pi ab}{N}}, (a, b = 0, \ldots, N-1)$, and $\mathbf{F}^H$ is an Inverse

Discrete Fourier Transform (IDFT) matrix, $\boldsymbol{\Phi}^{(\phi)} = \text{diag}\{[0, e^{\frac{j2\pi\phi}{N}}, \ldots, e^{\frac{j2\pi\phi(N-1)}{N}}]\}$ is the diagonal CFO matrix, with $\text{diag}\{\mathbf{x}\}$ denoting a diagonal matrix whose diagonal elements are entries of vector $\mathbf{x}$, $\mathbf{H} = \text{diag}\{[H(0), H(1), \ldots, H(N-1)]\}$ is the diagonal frequency-domain channel matrix, with $H(n)$ denoting the channel frequency response on the n-th subcarrier, and $\mathbf{z}_f$ is the $N \times 1$ Additive White Gaussian Noise (AWGN) vector.

Note that the CFO has a range of $[-0.5, 0.5)$ [15]. The ICI matrix caused by the CFO is written as $\mathbf{C}^{(\phi)} = \mathbf{F}\boldsymbol{\Phi}^{(\phi)}\mathbf{F}^H$. The ICI matrix $\mathbf{C}^{(\phi)}$ is circular, with each row equal to the previous one rotated by one element [16]. With the effect of the CFO, the received signal vector $\mathbf{y}^{(\phi)}$ can be written as

$$\mathbf{y}^{(\phi_f)} = \mathbf{C}^{(\phi)}\mathbf{H}\mathbf{s} + \mathbf{z}_f. \tag{2.4}$$

The frequency component on the n-th subcarrier is affected by the ICI from other $(N-1)$ subcarriers. In such a case, the orthogonality between subcarriers is destroyed by the CFO.

In a fast moving environment, the Doppler shift is determined by carrier frequency and velocity. This effect gives rise to time-varying channels in the time domain, and a frequency drift between subcarriers in the frequency domain. Thus, the orthogonality between subcarriers could not be maintained.

2.1.3 OFDMA

Recently, OFDMA, derived from OFDM, has attracted much research attention. It has been adopted by the Digital Video Broadcasting - Return Channel Terrestrial (DVB-RCT) [17], and the wireless local area networks IEEE 802.11 standards [6, 18].

Different from OFDM systems in which a single user occupies all subcarriers for the signal transmission, OFDMA systems allow multiple users to transmit symbols simultaneously using different orthogonal subcarriers. In OFDMA systems, a subset of subcarriers is assigned to each user, and the number of subcarriers for an individual user can be adaptively varied in each frame. Furthermore, OFDMA systems can provide a relatively lower PAPR than OFDM systems.

OFDMA has dynamic resource allocation, as it allows users to select their own subset of subcarriers according to channel conditions. Generally, there are three kinds of Carrier Assignment Scheme (CAS) available for users in OFDMA systems: subband CAS, interleaved CAS and generalized CAS, as shown in Fig. 2.1 [19].

- **Subband CAS**: several adjacent subcarriers are composed into a subblock to be allocated for an individual user.
- **Interleaved CAS**: the subcarriers of users are interleaved, and are uniformly spaced over the signal bandwidth at a distance from each other.

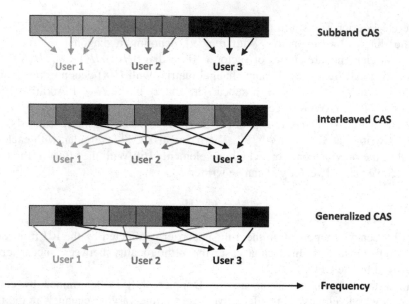

Fig. 2.1 CAS in OFDMA systems

- **Generalized CAS**: individual subcarrier can be assigned to any user, allowing
  dynamic resource allocation, since no specific rigid relationship between users
  and subcarriers exists.

Let N and K denote the total number of subcarriers and the number of users,
respectively. In OFDMA systems, one subcarrier can be assigned to one user only,
and could not be occupied by other users. Assume that all the subcarriers are used
for the transmission with no virtual or null subcarrier. Define $\boldsymbol{\Omega}_k(i)$ as the subset of
the number of subcarriers allocated to the k-th user ($k = 0, 1, \cdots, K - 1$) in the i-th
OFDMA block. It can be written as

$$\boldsymbol{\Omega}_k(i) = \{\Omega_k(0, i), \quad \Omega_k(1, i), \quad \cdots, \quad \Omega_k(N - 1, i)\}, \tag{2.5}$$

where $\Omega_k(n, i)$ denotes the indicator of subcarrier allocation for the k-th user on the
n-th subcarrier in the i-th OFDMA block, given as

$$\Omega_k(n, i) = \begin{cases} 1 \text{ if } \Omega_k(n, i) \text{ occupied by the } k\text{-th user} \\ 0 \text{ if } \Omega_k(n, i) \text{ not occupied by the } k\text{-th user} \end{cases}. \tag{2.6}$$

The union of subsets of subcarriers for K users satisfies

$$\boldsymbol{\Omega}_0(i) \bigcup \boldsymbol{\Omega}_1(i) \quad \cdots \quad \bigcup \boldsymbol{\Omega}_{K-1}(i) = \prod_N = \{N\}, \tag{2.7}$$

The intersection of subsets of subcarriers for K users satisfies

$$\boldsymbol{\Omega}_0(i) \quad \bigcap \quad \boldsymbol{\Omega}_1(i) \quad \cdots \quad \bigcap \quad \boldsymbol{\Omega}_{K-1}(i) = \emptyset, \qquad (2.8)$$

where $\emptyset$ denotes an empty set. Let $x_k(n, i)$ denote the symbol on the n-th subcarrier in the i-th block, and transmitted by the k-th user. After subcarrier allocation, the resulting symbol $s_k(n, i)$ in OFDMA systems is given by

$$s_k(n, i) = x_k(n, i) \cdot \Omega_k(n, i). \qquad (2.9)$$

The orthogonality between subcarriers provides the intrinsic protection against the Multiple-User Interference (MUI). Therefore, OFDMA inherits from OFDM, the ability to have a simple equalization scheme in the frequency domain. However, there are several technical challenges in OFDMA systems, some of which are frequency and timing synchronisation. Similar to OFDM, OFDMA is extremely sensitive to the CFO and STO [15, 20, 21]. As discussed previously, The CFO destroys the orthogonality between subcarriers and results in the ICI. The STO gives rise to either the ISI or phase shift in the received signals. In particular, multiple users are allowed to transmit simultaneously in OFDMA systems, posing more challenges for multi-CFO and multi-STO estimation. In Chap. 6, a joint algorithm for ICI mitigation and equalization is proposed in the OFDMA uplink.

2.2 MIMO OFDM Systems

To improve system capacity, multiple transmit and receive antennas are employed to establish multiple spatial branches, referred to as MIMO systems [22], as illustrated in Fig. 2.2. Compared to traditional SISO systems, MIMO systems can increase bandwidth efficiency, as multiple transmit and receive antennas operate on the same frequency band for the signal transmission. In order to combat frequency selective fading, OFDM is well suited for use in MIMO systems. Also, equalization can be simplified in the frequency domain for OFDM based systems. Therefore, MIMO OFDM systems have been adopted by the wireless local area networks IEEE 802.11 standards and LTE Advanced, respectively [6–8]. However, as the prorogation path between each transmit antenna and each receive antenna is independent, MIMO OFDM systems give rise to an additional spatial interference, known as the Co-Antenna Interference (CAI) or Co-Channel Interference (CCI) [23], which needs to be eliminated.

An MIMO OFDM system is considered, with K transmit and M receive antennas in the frequency selective fading environment, as illustrated in Fig. 2.4. The incoming serial data at the transmitter is divided into parallels. The IDFT/DFT pair allows

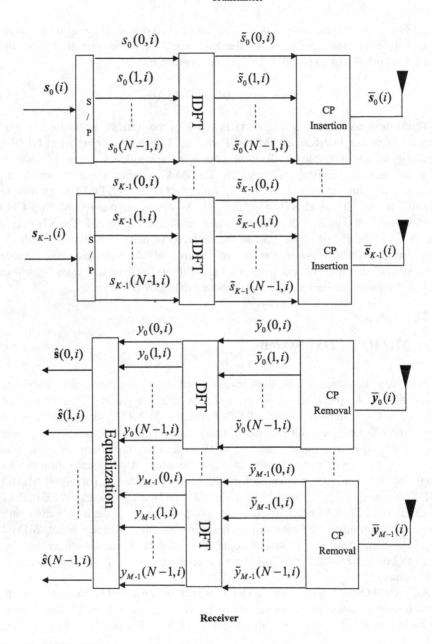

Fig. 2.2 MIMO OFDM systems block diagram

the signal to be transferred in between the frequency domain and the time domain. Let $s_k(n, i)$ denote the symbol on the n-th subcarrier ($n = 0, 1, \ldots, N - 1$) in the i-th block ($i = 0, 1, \ldots, N_s - 1$) transmitted by the k-th transmit antenna ($k = 0, 1, \ldots, K - 1$). Define $\mathbf{s}_k(i) = [s_k(0, i), s_k(1, i) \ldots, s_k(N - 1, i)]^T$ as the signal vector in the i-th block for the k-th transmit antenna. Details of the entire process of the data transfer from transmitter to receiver via channels are described in the following steps.

At the transmitter, the signal in the i-th OFDM block is first transformed to the time domain as $\tilde{\mathbf{s}}_k(i)$ by the IDFT as

$$\tilde{\mathbf{s}}_k(i) = \mathbf{F}^H \mathbf{s}_k(i), \tag{2.10}$$

where $\mathbf{F}$ is an $N \times N$ DFT matrix, with the (a, b)-th entry given by $\mathbf{F}(a, b) = \frac{1}{\sqrt{N}} e^{\frac{-j2\pi ab}{N}}$, $(a, b = 0, \ldots, N-1)$, and $\mathbf{F}^H$ is an IDFT matrix, with $\mathbf{F}^H = \mathbf{F}^{-1}$ since $\mathbf{F}$ is a unitary matrix [16]. Note that the computationally efficient IFFT/FFT pair may also be employed.

Assuming a total number of L channel paths, a Cyclic Prefix (CP) of length L_{CP}, at least $L_{CP} \geq L - 1$, is attached to each OFDM block $\tilde{\mathbf{s}}_k(i)$. The guard symbols consist of a copy of the last L_{CP} entries of each OFDM block. The insertion of a CP has two purposes: Inter Block Interference (IBI) avoidance and circular convolution between time-domain signal and Channel Impulse Response (CIR). With the CP insertion, the transmitted signal vector $\bar{\mathbf{s}}_k(i)$ can be given as

$$\bar{\mathbf{s}}_k(i) = \mathbf{T}_{CP} \tilde{\mathbf{s}}_k(i), \tag{2.11}$$

where $\mathbf{T}_{CP} = [\mathbf{I}_{CP}^T, \mathbf{I}_N^T]^T$ is the $(L_{CP} + N) \times N$ matrix, with $\mathbf{I}_N$ denoting an $N \times N$ identity matrix and $\mathbf{I}_{CP}$ denoting the last L_{CP} rows of $\mathbf{I}_N$.

The signal is then transmitted through the frequency selective fading channel, which is assumed to be constant for the duration of a frame consisting of a total number of N_s OFDM blocks. This is a convolution process as shown in Fig. 2.3. The received signal vector $\bar{\mathbf{y}}_m(i) = [\bar{y}_m(0, i), \bar{y}_m(1, i) \ldots, \bar{y}_m(N - 1, i)]^T$ at the m-th receive antenna ($m = 0, 1, \ldots, M - 1$) in the time domain can be written as

$$\bar{\mathbf{y}}_m(i) = \sum_{k=0}^{K-1} \bar{\mathbf{H}}_{m,k} \bar{\mathbf{s}}_k(i) + \bar{\mathbf{z}}_m(i), \tag{2.12}$$

where $\bar{\mathbf{H}}_{m,k}$ is the $(L_{CP} + N) \times (L_{CP} + N)$ convolutional channel matrix between the m-th receive antenna and the k-th transmit antenna, given as

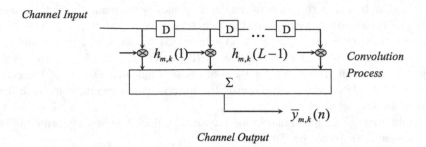

Fig. 2.3 Convolution process of signal and channel for MIMO OFDM systems

$$
\bar{\mathbf{H}}_{m,k} = \begin{bmatrix} h_{m,k}(0) & 0 & \cdots & \cdots & 0 \\ \vdots & h_{m,k}(0) & 0 & \cdots & 0 \\ h_{m,k}(L-1) & & \ddots & & \vdots \\ \vdots & \ddots & & \ddots & 0 \\ 0 & \cdots & h_{m,k}(L-1) & \cdots & h_{m,k}(0) \end{bmatrix}, \qquad (2.13)
$$

where $h_{m,k}(l)$ is the l-th ($l = 0, 1, \ldots, L-1$) channel path between the m-th receive antenna and the k-th transmit antenna, and $\bar{\mathbf{z}}_m(i)$ is the AWGN vector whose entries are Independent Identically Distributed (i.i.d.) complex Gaussian random variables with a zero mean and a variance of N_0 [22].

After the CP is removed, the received signal vector $\tilde{\mathbf{y}}_m(i)$ at the m-th receive antenna can be written as

$$
\tilde{\mathbf{y}}_m(i) = \mathbf{R}_{\text{CP}} \bar{\mathbf{y}}_m(i), \qquad (2.14)
$$

where $\mathbf{R}_{\text{CP}} = [\mathbf{0}_{N \times L_{\text{CP}}}, \mathbf{I}_N]$ is the $N \times (L_{\text{CP}} + N)$ matrix used to remove the CP, with $\mathbf{0}_{N \times L_{\text{CP}}}$ denoting the $N \times L_{\text{CP}}$ matrix filled with zeros.

The received signal is transformed to the frequency domain by applying the $N \times N$ DFT matrix to $\tilde{\mathbf{y}}_m(i)$ as

$$
\mathbf{y}_m(i) = \mathbf{F} \tilde{\mathbf{y}}_m(i). \qquad (2.15)
$$

The full circular convolution process between channel and signal can be achieved using the CP. The time-domain circular convolution can be transformed to a linear multiplication in the frequency domain by applying the IDFT/DFT pair, leading to simple equalization for the frequency selective fading environment [24]. The circulant matrix is a Toeplitz matrix where each row is equal to the previous one rotated by one element [16]. By using the IDFT/DFT pair, the circulant matrix can be diagonalized [16]. The resulting transceiver signal model in the frequency domain can be written as

$$y_m(i) = \sum_{k=0}^{K-1} \mathbf{H}_{m,k}\mathbf{s}_k(i) + \mathbf{z}_m(i), \tag{2.16}$$

where $\mathbf{H}_{m,k} = \mathbf{F}\tilde{\mathbf{H}}_{m,k}\mathbf{F}^H$ is the diagonal frequency-domain channel matrix, with $\tilde{\mathbf{H}}_{m,k} = \mathbf{R}_{\mathrm{CP}}\bar{\mathbf{H}}_{m,k}\mathbf{T}_{\mathrm{CP}}$ denoting the equivalent circulant channel matrix. The entry (n,n) in $\mathbf{H}_{m,k}$ is written as $H_{m,k}(n,n) = \sum_{l=0}^{L-1} h_{m,k}(l)e^{-\frac{j2\pi nl}{N}}$, and $\mathbf{z}_m(i) = \mathbf{F}\mathbf{R}_{\mathrm{CP}}\bar{\mathbf{z}}_m(i)$ is the frequency-domain noise vector. Note that the distribution statistics of the channel can be preserved by the DFT [22], if the CIR $h_{m,k}(l)$ is assumed to have the Rayleigh distributed magnitude and uniformly distributed phase. Also, the distribution of the white Gaussian noise samples can be preserved by the DFT.

Finally, as the frequency selective fading channel is divided into a number of flat fading channels, Frequency Domain Equalization (FDE) can be performed on each subcarrier to simplify the equalization process for MIMO OFDM systems, Define $\mathbf{s}(n,i) = [s_0(n,i), s_1(n,i), \ldots, s_{K-1}(n,i)]^T$ as the signal vector from K transmit antennas on the n-th subcarrier in the i-th block. The received signal vector $\mathbf{y}(n,i) = [y_0(n,i), y_1(n,i), \ldots, y_{M-1}(n,i)]^T$ in the frequency domain on the n-th subcarrier can be written as

$$\mathbf{y}(n,i) = \mathbf{H}(n)\mathbf{s}(n,i) + \mathbf{z}(n), \tag{2.17}$$

where $\mathbf{H}(n)$ is the $M \times K$ channel frequency response matrix on the n-th subcarrier, with $H_{m,k}(n)$ denoting the entry (m,k) in $\mathbf{H}(n)$, the channel frequency response between the m-th receive antenna and the k-th transmit antenna, and $\mathbf{z}(n)$ is the noise vector. The source data estimate $\hat{\mathbf{s}}(n,i)$ can be performed by either Zero Forcing (ZF) or Minimum Mean Square Error (MMSE) based equalization on the received signal on the n-th subcarrier as

$$\hat{\mathbf{s}}(n,i) = \mathbf{G}(n,i)\mathbf{y}(n,i), \tag{2.18}$$

where $\mathbf{G}(n,i)$ is the weighting matrix that can have either ZF or MMSE equalization criterion. The details of these equalization schemes are provided in the next chapter.

2.3 CoMP Transmission

The inter-cell interference is a major bottleneck for achieving very high data rates in wireless communication systems [23]. Previously, adjacent cells were operated on different frequencies to effectively reduce the inter-cell interference [25]. Recent research trends hint that millimeter (mm) wave communication will be the key component in 5G cellular systems. The short coverage of mm wave bands results in small cells. As the demand for high mobile data rates grows, higher spectral efficiency of cellular networks is needed, with full frequency reuse [26, 27]. One of best ways to manage interference is to allow each BS to connect to each other

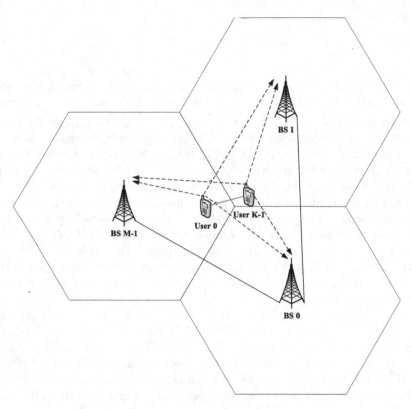

Fig. 2.4 Wireless CoMP system model with K users and M BSs

through a backhaul link. The BSs could exchange messages and jointly process received multiple users' signals on the same frequency band. This structure, called CoMP, has been adopted by LTE-Advanced [28]. CoMP transmission explores the interference between cells, which is different from other existing methods by treating them as noise [29, 30]. These features of CoMP systems are important for users at the cell-edge to have effective communication through the BSs. Also, CoMP is a cost-effective structure, as it requires little change to the current system. The general CoMP system diagram is illustrated in Fig. 2.4.

CoMP transmission has some significant challenges, which are summarized as follows.

- Multi-CFO estimation: As each BS or user has its own oscillator, there are multiple CFOs in the OFDM based CoMP transmission. The multi-CFO estimation becomes much difficult so that conventional frequency synchronization schemes for a single CFO are not suitable in the scenario. Compared to OFDMA systems, multi-CFO estimation in the CoMP transmission is much more challenging. There are two reasons. Firstly, the CoMP system allows users to transmit signals simultaneously by using all shared subcarriers, while multiple users could not share the same subcarriers in OFDMA systems. In the presence of multiple CFOs, one user's signal power is leaked into other users in OFDMA systems. The effect disappears with correct CFO compensation. By using this property, a number of multi-CFO estimation approaches were proposed for OFDMA systems [20, 31]. However, in the CoMP transmission, different users' signals interfere with each other at different BSs. This interference could not be removed, even in the case of no CFO. These multi-CFO estimation approaches used in OFDMA systems will significantly degrade the performance in the CoMP transmission, and therefore, are not suitable. Secondly, as frequency synchronization of all BSs and users is required, the number of CFOs linearly increases with the increasing number of BSs and users in the CoMP transmission. While in OFDMA systems, the number of CFOs only increases with the number of users. Thus, the number of CFOs in CoMP systems is larger than that in OFDMA systems.

- Multi-cell channel estimation and equalization: On the one hand, the central station in the CoMP transmission requires additional time to collect all received signals from multiple BSs for joint processing. This delay might cause a serious situation in time-limited communications. Thus, low-complexity multi-cell channel estimation and equalization are important and challenging in the CoMP transmission [32]. On the other hand, bandwidth resource is very scarce in wireless communication systems. Traditionally, training signals are commonly used for channel estimation. However, transmitting training signals reduces spectral efficiency. Therefore, there is a trade-off between complexity and spectral efficiency, when designing channel estimation and equalization schemes for the CoMP transmission.

In Chap. 5, a low-complexity multi-CFO estimation method and an ICA based equalization scheme are presented for CoMP systems to well deal with the challenges above.

2.4 CA Technology

CA, as one of key features in LTE-Advanced [28, 33], allows several smaller component carriers (spectrum chunk) to be aggregated. Thus, high data rates can be achieved in the CA transmission, by deploying extended bandwidth for the concurrent transmission. The main purposes for introducing the CA are listed as follows.

- High data rates: Up to five component carriers can be allowed for the aggregation in the both uplink and downlink. As the bandwidth is up to 20 MHz for each component carrier, a maximum of 100 MHz is achieved in the supported bandwidth for five component carriers in total. The peak target data rates are in excess of 1 Gbps in the downlink and 500 Mbps in the uplink, respectively.

- Configuration flexibility: It is possible to have asymmetric configurations for the CA at user and BS. Also, the number of component carriers could be used differently in the downlink and uplink.
- Frequency flexibility: As the locations of users are different, they can select component carriers based on several conditions, such as CSI. This can provide great frequency flexibility. This flexibility also allows for efficient support in power control and component carrier allocation.
- Interference management: the CA can be used as a promising inter-cell interference coordinator, and is dependent on many factors: the relative locations of BSs, traffic situation, the mutual interference coupling and so on. These factors are configured to optimize system performance.

In terms of frequency location, there are three different aggregation scenarios, as shown in Fig. 2.5 [34]. These aggregation scenarios are described as below.

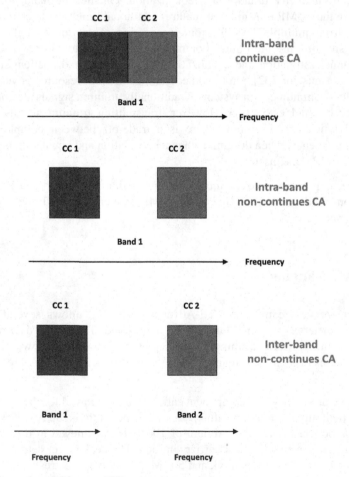

Fig. 2.5 A number of CA types (CC: component carrier)

- Intra-band aggregation with contiguous carriers: A number of continuous component carriers are aggregated within the same bandwidth.
- Intra-band aggregation with non-contiguous carriers: A number of separated component carriers, belonging to the same bandwidth, are aggregated.
- Inter-band aggregation with non-contiguous carriers: A number of non-adjacent component carriers over different bands are aggregated. The exploitation of fragmented spectrums is utilized to enhance frequency flexibility, as the idle bandwidth available can be employed.

For the case of intra-band aggregation with contiguous carriers, only a single transmit chain is used, because the aggregated carriers are contiguous. While there are multiple transmit chains in the non-contiguous carriers. The CA type selection, either contiguous or non-contiguous, depends on the trade-off between cost, complexity and the range of transmission bandwidth. One of targets for LTE-Advanced is expected to provide improvement in cell-edge spectral efficiency [28]. To meet the target, the inter-band non-contiguous CA can be effectively employed in CoMP systems to improve cell-edge throughput, since other idle bandwidths can be combined for the concurrent transmission. However, extra multiple CFOs occur since multiple LOs are used to support the non-continuous carriers on different frequency bands. In Chap. 6, a solution to the multi-CFO problem is proposed for CA based CoMP OFDMA systems by using a semi-blind ICA based joint ICI mitigation and equalization scheme.

References

1. R. Schober. Noncoherent space-time equalization. *IEEE Transactions on Wireless Communications*, 2(3):537–548, May 2003.
2. L. Sarperi, X. Zhu, and A. K. Nandi. Reduced complexity blind layered space-time equalization for MIMO OFDM systems. In *Proc.IEEE International Symposium on Personal Indoor and Mobile Radio Communications (PIMRC)*, Berlin, Germany, Sep. 2005.
3. X. Zhu and R. D. Murch. Layered space-time equalization for wireless MIMO systems. *IEEE Transactions on Wireless Communications*, 2(6):1189–1203, Nov. 2003.
4. S.Weinstein and P. Ebert. Data transmission by frequency-division multiplexing using the discrete fourier transform. *IEEE Transactions on Communications*, 19(5):628–634, Oct. 1971.
5. Z. Wang and G. B. Giannakis. Wireless multicarrier communications. *IEEE Signal Processing Magazine*, 17(3):29–48, May 2000.
6. Ieee standard for information technology-telecommunications and information exchange between systems-local and metropolitan area networks-specific requirements part 11: Wireless lan medium access control (mac) and physical layer (phy) specifications, Mar. 2012.
7. K. Fazel and S. Kaiser. *Multi-carrier and spread spectrum systems: from OFDM and MC-CDMA to LTE and WiMAX*. John Wiley & Sons, New York, USA, second edition, 2008.
8. A. Technologies. *LTE and the Evolution to 4G Wireless: Design and Measurement Challenges*. John Wiley & Sons, New York, USA, 2009.
9. C. J. Hansen. WiGiG: Multi-gigabit wireless communications in the 60 Ghz band. *IEEE Wireless Communications Magazine*, 18(6):6–7, Dec. 2011.
10. Ieee draft standard for local and metropolitan area networks - specific requirements - part 11: Wireless lan medium access control (mac) and physical layer (phy) specifications amendment 3: Enhancements for very high throughput in the 60 ghz band, Jul. 2012.

11. A. Georgakopoulos, D. Karvounas, and K. Tsagkaris. 5G on the horizon: key challenges for the radio-access network. *IEEE Vehicular Technology Magazine*, 8(3):47–53, Jul. 2013.
12. C. Edwards. 5G searches for formula to shake off Shannon. *Engineering & Technology*, 8(8):82–85, Oct. 2013.
13. J. A. C. Bingham. Multi-carrier modulation for data transmission: an idea whose time has come. *IEEE Communications Magazine*, 28(5):17–25, May 1990.
14. P. P. Vaidyanathan. *Multi-Rate System and Filter Banks*. Prentice Hall, 1993.
15. Y. S. Cho, J. Kim, W. Y. Yang, and C. G. Kang. *MIMO-OFDM Wireless Communications with MATLAB*. Wiley, Singapore, 2010.
16. C. D. Meyer. *Matrix Analysis and Applied Linear Algebra*. Society for Industrial and Applied Mathematics, Philadelphia, U.S.A., 2000.
17. Interaction channel for digital terrestrial television (rct) incorporating multiple access ofdm,, Mar. 2001.
18. Ieee standard for local and metropolitan area networks, part 16: Air interface for fixed and mobile broadband wireless access systems amendment2: Physical and medium access control layers for combined fixed and mobile operation in licensed bands, 2005.
19. M. Morelli, C. C. J. Kuo, and M. O. Pun. Synchronization techniques for orthogonal frequency division multiple access (OFDMA): a tutorial review. *Proceedings of the IEEE*, 95(7):1394–1427, Jul. 2007.
20. M. Movahhedian, Y. Ma, and R. Tafazolli. Blind CFO estimation for linearly precoded OFDMA uplink. *IEEE Transactions on Signal Processing*, 58(9):4698–4710, Sep. 2010.
21. S. Manohar, D. Sreedhar, V. Tikiya, and A. Chockalingam. Cancellation of multiuser interference due to carrier frequency offsets in uplink OFDMA. *IEEE Transactions on Wireless Communications*, 6(7):2560–2571, Jul. 2007.
22. A. J. Paulraj, R. Nabar, and D. Gore. *Introduction to Space-Time Wireless Communications*. Cambridge University Press, Cambridge, U.K., 2003.
23. A. Goldsmith. *Wireless Communications*. Cambridge University Press, London, U.K., 2005.
24. S. J. Orfanidis. *Introduction to Signal Processing*. Prentice Hall, Upper Saddle River, U.S.A., 1996.
25. T.D. Novlan, R. K. Ganti, A. Ghosh, and J. G. Andrews. Analytical evaluation of fractional frequency reuse for OFDMA cellular networks. *IEEE Transactions on Wireless Communications*, 10(12):4294–4305, Dec. 2011.
26. A. Sklavos, T. Weber, E. Costa, H. Haas, and E. Schulz. Joint detection in multi-antenna and multi-user OFDM systems. *Multi-Carrier Spread Spectrum and Related Topics*, pages 191–198, May 2002.
27. S. Shamai, O. Somekh, and B. Zaidel. Multi-cell communications: a new look at interference. *IEEE Journal on Selected Areas in Communications*, 28(9):1380–1408, Dec. 2010.
28. 3gpp technical report 36.814 version 9.0.0, further advancements for e-utra physical layer aspects, Mar. 2010.
29. J. Andrews. Interference cancellation for cellular systems: a contemporary overview. *IEEE Transaction on Wireless Communications*, 12(2):19–29, Apr. 2005.
30. P. Marsch, S. Khattak, and G. Fettweis. A framework for determining realistic capacity bounds for distributed antenna systems. In *Proc. IEEE Information Theory Workshop*, pages 571–575, Chengdu, China, Oct. 2006.
31. P. Sun and L. Zhang. Low complexity pilot aided frequency synchronization for OFDMA uplink transmission. *IEEE Transaction on Wireless Communications*, 8(7):3758–3769, Jul. 2009.
32. J. Hoydis, M. Kobayashi, and M. Debbah. Optimal channel training in uplink network MIMO systems. *IEEE Transactions on Signal Processing*, 59(6):2824–2834, 2011.
33. Z. Shen, A. Papasakellariou, J. Montojo, D. Gerstenberger, and F. Xu. Overview of 3GPP LTE-Advanced carrier aggregation for 4G wireless communications. *IEEE Communications Magazine*, 50(2):122–130, Feb. 2012.
34. K. I. Pedersen, F. Frederiksen, C. Rosa, L. G. U. Garcia H. Nguyen, and Y. Wang. Carrier aggregation for LTE Advanced: functionality and performance aspects. *IEEE Communications Magazine*, 49(6):89–95, Jun. 2011.

Chapter 3
CFO Estimation and Channel Equalization for OFDM Based Wireless Systems

3.1 CFO Estimation Methods

In this section, a number of preamble based and blind or semi-blind CFO estimation methods are described. A number of preamble based methods are reviewed, including traditional Moose's CFO estimation, powerful Maximum Likelihood (ML) [1] based CFO estimation and latest Constant Amplitude Zero Auto-Correlation (CAZAC) sequences [2–4] based CFO estimation approaches. Also, a number of blind or semi-blind methods are presented, including diagonality criterion based blind CFO estimation, CP based blind CFO estimation, precoding based semi-blind CFO estimation and pilot based semi-blind CFO estimation approaches.

3.1.1 Preamble Based CFO Estimation

In [5], the effect of CFO on the channel estimation performance was analyzed. It is revealed that the Mean Square Error (MSE) performance of channel estimation in the presence of CFO could not be improved very much by increasing the number of training symbols, because the ICI exists between subcarriers. Thus, it is important to perform correct CFO estimation and compensation prior to channel estimation and equalization for the accurate recovery of the transmitted data. Generally, the training signal for CFO estimation is transmitted during the period of preamble, where the source signal is not applicable. This kind of CFO estimation can provide very good performance, as a large number of well-designed training sequences can be transmitted. There are a number of commonly used and recently proposed CFO estimation methods in the literature, which are introduced below.

© The Author(s) 2015
Y. Jiang et al., *Semi-Blind Carrier Frequency Offset Estimation and Channel Equalization*, SpringerBriefs in Electrical and Computer Engineering,
DOI 10.1007/978-3-319-24984-1_3

3.1.1.1 Moose's CFO Estimation

Moose's CFO estimation is a traditional method [6] that allows successive OFDM blocks to be identical. A phase shift caused by the CFO between successive blocks is used for CFO estimation. By using the correlation between successive blocks in the frequency domain, the CFO $\hat{\phi}$ can be estimated as

$$\hat{\phi} = \frac{1}{2\pi}\tan^{-1}\left\{\frac{\sum_{n=0}^{N-1}\Im m[y(n,1)y^*(n,2)]}{\sum_{n=0}^{N-1}\Re e[y(n,1)y^*(n,2)]}\right\}, \tag{3.1}$$

where $\Im m\{\cdot\}$ and $\Re e\{\cdot\}$ denote the imaginary part and the real part of a complex number, respectively, and $y(n,1)$ and $y(n,2)$ are the received symbols on the n-th subcarrier in the first and second blocks, respectively.

Moose's CFO estimation method [6] is simple, as a linear Least-Square (LS) based algorithm is obtained. However, a large number of symbols are required to obtain a good performance. Also, this method is sensitive to CFO variations, and offers a poor performance for a large CFO value. Thereby, the CFO estimation accuracy of Moose's method is not guaranteed for OFDM systems. In [7], Moose's CFO estimation method is extended to a joint Symbol Time Offset (STO) and CFO estimation approach. In order to overcome the error floor in the MSE performance, a recursive LS based iterative CFO estimation approach is developed as in [8], which is robust against channel and CFO variations.

3.1.1.2 ML Based CFO Estimation

The ML based CFO estimation method [1] is one of most widely used approaches. This method requires a search to be performed over the possible range of CFO values. Thus, this kind of CFO estimation can be viewed as an optimization problem aimed at maximizing an objective function. In [1], a multiuser OFDM system is considered, each user equipped with a single antenna. Define $\tilde{\phi}_k$ as the trial CFO value for the k-th user. Correspondingly, the k-th user's CFO matrix with the trial value can be written as $\Phi^{(\tilde{\phi}_k)} = \text{diag}\{[1, e^{\frac{j2\pi\tilde{\phi}_k}{N}}, \ldots, e^{\frac{2\pi\tilde{\phi}_k(N-1)}{N}}]\}$. Let $\mathbf{S}_k$ denote the $N \times N$ circulant matrix transmitted by the k-th user, with the first column $[s_k(0), s_k(1), \ldots, s_k(N-1)]^T$. Incorporating the effect of CFO, the $N \times (N \times K)$ transmitted signals matrix $\mathbf{X}^{(\tilde{\phi})}$ from all K users is written as $\mathbf{X}^{(\tilde{\phi})} = [\Phi^{(\tilde{\phi}_0)}\mathbf{S}_0, \Phi^{(\tilde{\phi}_1)}\mathbf{S}_1, \ldots, \Phi^{(\tilde{\phi}_{K-1})}\mathbf{S}_{K-1}]$, with $\tilde{\phi} = \{\tilde{\phi}_0, \tilde{\phi}_1, \ldots, \tilde{\phi}_{K-1}\}$ denoting the trial CFOs from all K users. The ML based estimate of multiple CFOs $\hat{\phi} = \{\hat{\phi}_0, \hat{\phi}_1, \ldots, \hat{\phi}_{K-1}\}$ for all K users can be given as

$$\hat{\phi} = \arg\max_{\tilde{\phi}}\left\{\text{tr}\left[\mathbf{Y}^H\mathbf{X}^{(\tilde{\phi})}([\mathbf{X}^{(\tilde{\phi})}]^H\mathbf{X}^{(\tilde{\phi})})^{-1}[\mathbf{X}^{(\tilde{\phi})}]^H\mathbf{Y}\right]\right\}, \tag{3.2}$$

where tr($\cdot$) denotes the trace of a matrix, $\mathbf{Y} = [\mathbf{y}_0, \mathbf{y}_1, \ldots, \mathbf{y}_{M-1}]$ is the $N \times M$ received signals matrix from all M receive antennas, with $\mathbf{y}_m$ denoting the received signal vector at the m-th receive antenna.

The ML based CFO estimation method is a powerful estimator, and has high accuracy. However, the complexity of this search grows exponentially with the number of users, and hence it is not practical to implement for multi-CFO estimation.

3.1.1.3 CAZAC Sequences Based CFO Estimation

In [2–4], CAZAC sequences were employed for multi-CFO estimation, using their properties of constant amplitude elements and zero auto-correlation for any nonzero circular shift. Let $s(n) = e^{j2\pi\theta_n}$ denote the n-th symbol in the length-N CAZAC sequences, with θ_n being the angle. The auto-correlation between CAZAC sequences is expressed as

$$\sum_{n=0}^{N-1}[s(n)s^*(n \ominus l)] = \begin{cases} N & l = 0 \\ 0 & l \neq 0 \end{cases}, \tag{3.3}$$

where $\ominus$ denotes the circular subtraction and $l = 0, 1, \ldots, N-1$. Let $\mathbf{S}_i$ denote the i-th ($i = 0, 1, \ldots, K-1$) user's $N \times N$ circulant matrix generated by the length-N CAZAC sequences, where the first column is equal to $[s_i(0), s_i(1), \ldots, s_i(N-1)]^T$. The auto-correlation and cross-correlation of CAZAC sequences between users i and j ($i, j = 0, 1, \ldots, K-1$) are written as

$$\mathbf{S}_i^H \mathbf{S}_j = \begin{cases} N\mathbf{I}_N & i = j \\ N\mathbf{I}_N^{(\sigma_j - \sigma_i)} & i \neq j \end{cases}, \tag{3.4}$$

where $\mathbf{I}_N^{(\sigma_j - \sigma_i)}$ denotes a matrix resulting from the identity matrix, with the location of 1 cyclically shifted to the right by $(\sigma_j - \sigma_i)$ positions.

As multiple CFOs could destroy the orthogonality between different users' CAZAC sequences, the difference between CFOs has impact on the estimation performance. In order to reduce the effect, a number of CAZAC sequences need to be selected. Actually, there are three classes for CAZAC sequences: Frank and Zadoff sequences [9], Chu sequences [10] and S&H sequences [11]. There are a finite number of Frank and Zadoff sequences or Chu sequences, while an infinite number of S&H sequences. The optimal CAZAC sequences selection can be obtained from the limited number of Frank and Zadoff sequences or Chu sequences. However, this requires solving a multi-dimensional optimization problem. The high complexity limits its practical use. Even if the multi-dimensional search can be reduced to a number of one-dimensional searches by regarding other users' CFOs as noise, an inevitable estimation error happens at a high Signal-to-Noise Ratio (SNR), giving rise to a large error floor in MSE and BER performances. Thus, a large number of symbols are required to reduce the effect of error floor.

In order to overcome these drawbacks, the CAZAC sequences based CFO estimation method was extended to CoMP systems in CAZAC and [4]. By incorporating iterative interference cancellation, the multi-CFO estimation scheme is robust against CFOs variations, and provides a good performance.

3.1.2 Blind and Semi-Blind CFO Estimation

Blind or semi-blind CFO estimation methods are based on using little or no prior knowledge at the receiver. Since transmitting a large number of training signals is not required, blind and semi-blind approaches have high spectral efficiency. However, most of them have high complexity, particularly for the case of multiple CFOs. In this section, some state-of-the-art blind and semi-blind CFO estimation methods are presented, such as diagonality criterion based blind CFO estimation, CP based blind CFO estimation, precoding based semi-blind CFO estimation and pilot based semi-blind CFO estimation approaches.

3.1.2.1 Diagonality Criterion Based Blind CFO Estimation

In [12], a diagonality criterion based blind CFO estimation method was proposed, requiring no prior knowledge of the transmitted data and channel. In CFO-free OFDM systems, the auto-correlation of the received signal in the frequency domain is an approximation to a diagonal structure, where the off-diagonal elements are close to zero. However, the presence of CFO destroys the orthogonality between subcarriers, leading to a non-diagonal structure. Based on the property, the received signals are passed to the CFO compensator with the trial value first. The CFO estimation is then performed by minimizing the power of non-diagonal elements. The auto-correlation of the received signal is forced into a diagonal structure. Define ϕ and $\tilde{\phi}$ as the real and the trial CFO values, respectively. Let $\mathbf{C}(\phi - \tilde{\phi})$ denote the trial CFO compensated ICI matrix. Define $\mathbf{Q}^{(\phi - \tilde{\phi})}$ as the auto-correlation of the received signal with the trial CFO compensation, written as

$$\mathbf{Q}^{(\phi - \tilde{\phi})} = \mathbf{F}_N \mathbf{C}^*(\phi - \tilde{\phi})\bar{\mathbf{H}}\bar{\mathbf{H}}^H \mathbf{C}(\phi - \tilde{\phi})\mathbf{F}_N + \sigma^2 \mathbf{I}_N, \tag{3.5}$$

with $\mathbf{F}_N$ and $\bar{\mathbf{H}}$ denoting the $N \times N$ DFT matrix and the circulant channel matrix, respectively. The diagonality criterion based CFO estimation method can be given by

$$\hat{\phi} = \underset{\tilde{\phi} \in [-0.5, 0.5]}{\arg \min} \ ||\mathbf{Q}^{(\phi - \tilde{\phi})} \odot (\mathbf{1}_N - \mathbf{I}_N)||_F^2, \tag{3.6}$$

where $\odot$ and $|| \cdot ||_F^2$ denote the Hadamard product and the Frobenius norm, respectively, and $\mathbf{1}_N$ is an $N \times N$ all-one matrix.

Under the assumption of an infinite number of OFDM blocks, $\mathbf{Q}^{(\phi-\tilde{\phi})}$ approaches a diagonal matrix with the correct CFO compensation. Otherwise, the off-diagonal elements in $\mathbf{Q}^{(\phi-\tilde{\phi})}$ are not close to zero any more. A large number of blocks are therefore required to achieve a steady status. Due to the bias CFO estimation, there is an inevitable high error floor at the range of high SNRs.

3.1.2.2   CP Based Blind CFO Estimation

In order to avoid the IBI, the CP is introduced in OFDM systems. In [13] and [14], the CP was also used for blind CFO estimation. With a total number of N subcarriers, a phase difference of $2\pi N\phi/N = 2\pi\phi$ is induced by the CFO between the CP and the rest of an OFDM block. Therefore, the CP can be explored for blind CFO estimation, given as

$$\hat{\phi} = \frac{1}{2\pi} \arg\left\{ \sum_{n=-L_{CP}}^{-1} y(n)y^*(n+N) \right\}, \qquad (3.7)$$

where $\arg\{\cdot\}$ denotes the angle of a complex, $y(n)$ is the received symbol on the n-th subcarrier and L_{CP} denotes the length of the CP.

Actually, $y(n)y^*(n+N)$ becomes real only when there is no frequency offset, while it becomes imaginary as long as the CFO exists. Thus, the imaginary part can be used for CFO estimation. Since a closed-form solution is achieved, the CP based blind CFO estimation method has lower complexity, compared to other blind approaches. However, it is highly sensitive to the channel frequency selectivity. Thus, there is a high error floor at high SNRs. This effect can be reduced by increasing the number of symbols in the CP. However, it is not spectrally efficient.

3.1.2.3   Precoding Based Semi-Blind CFO Estimation

A novel semi-blind CFO estimation method, based on a linear precoding, was proposed in [15]. The OFDM block is divided into two subblocks, with each superimposed by a number of reference data sequences via a non-redundant precoding process. This precoding is a general class proposed by Lin and Petropulu in [16]. The correlation between subblocks induced by precoding is used to perform CFO estimation. The CFO can be estimated by

$$\hat{\phi} = \frac{N \arg\{\mu\}}{2\pi N_g}, \qquad (3.8)$$

where $N_g = N + L_{CP}$ is the length of the OFDM block, with N and L_{CP} denoting the number of subcarriers N and the length of the CP, respectively, μ is the correlation between subblocks as defined in [15].

Similar to other blind CFO estimation methods, a large number of blocks are required to achieve a good MSE performance. Also, there is an error floor from a SNR = 15 dB and above, as shown in [15].

3.1.2.4   Pilot Based Semi-Blind CFO Estimation

A number of semi-blind CFO estimation methods can be performed by a small number of known pilots which are transmitted together with the source symbols. This is different from training signals for CFO estimation during the period of preamble, when the source symbols are not available to be transmitted. Typically, the insertion of pilots has multiple purposes, such as channel estimation, frequency synchronization, timing synchronization and so on.

A semi-blind CFO estimation approach and a channel estimation method were proposed in [17] for MIMO OFDM systems, where one pilot OFDM block is exploited for CFO estimation and ambiguity elimination in the subspace based channel estimation.

In [18], a small number of pilot symbols inserted on two consecutive OFDMA blocks were employed for multi-CFO estimation. These pilots can also be used to track the CFO due to the Doppler shift. In order to resolve a multi-dimensional search problem for multiple CFOs, a number of one-dimensional searches can be performed by treating other users' CFOs as small. This assumption could cause a relatively inferior performance that can be improved by an iterative joint CFO estimation and compensation scheme. The one-dimensional search based CFO estimation method for the k-th user can be written as

$$\hat{\phi}_k = \arg\min_{\tilde{\phi}_k} \left\| \boldsymbol{\Omega}_k(2)\left[\boldsymbol{\Gamma}^{(\tilde{\phi}_k)}\right]^*\left[\mathbf{C}_k^{(\tilde{\phi}_k)}\right]^{-1}\mathbf{y2}^{(g)} - \boldsymbol{\Omega}_k(1)\left[\mathbf{C}_k^{(\tilde{\phi}_k)}\right]^{-1}\mathbf{y1}^{(g)}\right\|_F^2, \quad (3.9)$$

where $\boldsymbol{\Gamma}^{(\tilde{\phi}_k)}$ denotes the common phase error in the second OFDMA block, $\boldsymbol{\Omega}_k(i)$ is the subcarrier allocation indicator in the i-th block for the k-th user, $\mathbf{y1}^{(g)}$ and $\mathbf{y2}^{(g)}$ denote the received signals in the first and second OFDMA blocks, respectively, with g being the g-th iteration of CFO estimation. Performed iteratively, the interference from other users can be removed progressively until the orthogonality between subcarriers is restored.

3.2   Channel Estimation and Equalization

Channel estimation and equalization can be classified into training based and blind or semi-blind approaches, respectively. For training based channel estimation and equalization, a large number of training symbols are required, while they are not required for blind or semi-blind equalization. Therefore, blind or semi-blind equalization has higher spectral efficiency than training based channel estimation and equalization.

3.2.1   *Training Based Channel Estimation and Equalization*

In order to estimate the source signals, training signals are first transmitted for channel estimation. Then, equalization is performed using the CSI estimated by the training signals.

3.2.1.1   Channel Estimation

In this section, three widely used channel estimation schemes are discussed: LS based channel estimation, MMSE based channel estimation [19, 20] and channel interpolation [21] methods.

LS Based Channel Estimation

The LS based method has low complexity, and thus has been widely used for channel estimation. Define $\mathbf{h} = [h_o, \dots, h_{L-1}, h_L, \dots, h_{N-1}]^T$ as the channel vector, of which elements are assumed to be Gaussian variables and independent to each other. If there are a total number of L channel paths, then $h_L = \dots = h_{N-1} = 0$. The channel response energy is normalized to unity as $\sum_{l=0}^{N-1} \mathrm{E}\{h_l^2\} = 1$. In SISO systems, the received signal vector $\mathbf{y}(i) = [y(0, i), y(1, i), \dots, y(N-1, i)]^T$ in the i-th block can be written as

$$\mathbf{y}(i) = \mathbf{X}(i)\mathbf{H} + \mathbf{z}(i), \tag{3.10}$$

where $\mathbf{X}(i) = \mathrm{diag}\{[x(0, i), x(1, i), \dots, x(N-1, i)]^T\}$ is the $N \times N$ diagonal training matrix, $\mathbf{H} = \sqrt{N}\mathbf{F}\mathbf{h}$ is the channel frequency response vector on N subcarriers, and $\mathbf{z}(i)$ is the AWGN vector.

Assuming a total number of N_s blocks as training, the LS based channel estimate $\hat{\mathbf{H}}_{\mathrm{LS}}$ can be given by

$$\hat{\mathbf{H}}_{\mathrm{LS}} = \frac{1}{N_s} \sum_{i=0}^{N_s-1} \mathbf{X}^{-1}(i)\mathbf{y}(i). \tag{3.11}$$

Substituting Eqs. (3.10) into (3.11) yields

$$\hat{\mathbf{H}}_{\mathrm{LS}} = \frac{1}{N_s} \sum_{i=0}^{N_s-1} \left[\mathbf{H} + \mathbf{X}^{-1}(i)\mathbf{z}(i)\right]. \tag{3.12}$$

The term $\mathbf{X}^{-1}(i)\mathbf{z}(i)$ may be subject to noise enhancement, especially when the channel is in a deep null.

For MIMO OFDM systems, channel estimation can be decoupled into a number of independent SISO OFDM channel estimations, if the training symbols of transmit antennas are orthogonal to each other [22]. It means that a different subset of subcarriers is used by each transmit antenna for the training symbols transmission.

MMSE Based Channel Estimation

MMSE usually outperforms LS in channel estimation, as it can suppress the noise enhancement for known channel characteristics [23]. The MMSE based channel estimate $\hat{\mathbf{H}}_{MMSE}$ is performed by minimizing the following MSE

$$\min \ E\{||\hat{\mathbf{H}}_{MMSE} - \mathbf{H}||^2\}, \tag{3.13}$$

where $E\{\cdot\}$ denotes the expectation. By using the LS based channel estimate, the MMSE based channel estimation method can be given by [24]

$$\hat{\mathbf{H}}_{MMSE} = \mathbf{R}_{HH}(\mathbf{R}_{HH} + \sigma_z^2 \mathbf{I}_N)^{-1}\hat{\mathbf{H}}_{LS}, \tag{3.14}$$

where $\mathbf{R}_{HH} = E\{\mathbf{H}\mathbf{H}^H\}$ is the auto-correlation of channel frequency response.

Channel Interpolation

Channel interpolation [21] can be employed to refine the channel estimate performed by the LS based method. The correlation between adjacent subcarriers is used to correct some incorrect channel estimates for a few subcarriers. By using the LS based channel estimation $\hat{\mathbf{H}}_{LS}$ in Eq. (3.12), the time-domain channel estimation $\tilde{\mathbf{h}}$ is given as [21]

$$\tilde{\mathbf{h}} = \frac{1}{\sqrt{N}}\mathbf{F}_{N \times L}^{+}\hat{\mathbf{H}}_{LS}, \tag{3.15}$$

where $(\cdot)^+$ denotes the pseudo-inverse, $\mathbf{F}_{N \times L}$ is the $N \times L$ DFT matrix, with entry (a, b) given by $\mathbf{F}_{N \times L}(a, b) = \frac{1}{\sqrt{N}}e^{\frac{-j2\pi ab}{N}}$ $(a = 0, 1, \ldots, N-1; b = 0, 1, \ldots, L-1)$. The channel information for all subcarriers is used so that $\tilde{\mathbf{h}}$ is not influenced by a few errors on some subcarriers. After inserting $(N - L)$ zeros at the end of $\tilde{\mathbf{h}}$, the $N \times 1$ vector $\hat{\mathbf{h}}$ is written as $\hat{\mathbf{h}} = [\tilde{\mathbf{h}}, 0, \ldots, 0]^T$. The refined channel estimate $\hat{\mathbf{H}}_{CI}$ in the frequency domain can be given as

$$\hat{\mathbf{H}}_{CI} = \sqrt{N}\mathbf{F}\hat{\mathbf{h}}. \tag{3.16}$$

3.2.1.2 Equalization

By using the estimate of the CSI, equalization is performed to recover the transmitted signal on each subcarrier in OFDM based wireless communication systems. In this subsection, a number of equalization schemes are presented here: ZF and MMSE based equalization, Vertical Bell Laboratories Layered Space-Time (V-BLAST) based equalization and ML based equalization approaches.

ZF Based Equalization

In order to solve the problem of the CCI and ISI in OFDM based wireless communication systems, the ZF based equalizer applies the inverse of the channel frequency response to the received signal in the frequency domain. Due to the simplicity, it has been widely used in wireless communications systems. However, the noise power might be enhanced after the process, particularly for the subcarriers in the deep fading. When such a consequence arises, the received signal energy may be weak at some frequencies. Define $\mathbf{y}(n, i)$ as the received signal vector on the n-th subcarrier in the i-th block for OFDM based systems, the ZF based equalization method is performed on the n-th subcarrier as

$$\hat{\mathbf{s}}(n, i) = \mathbf{G}_{ZF}(n)\mathbf{y}(n, i), \tag{3.17}$$

where $\hat{\mathbf{s}}(n, i)$ denotes the equalized signal vector, and $\mathbf{G}_{ZF}(n)$ is the ZF equalizer on the n-th subcarrier, given by

$$\mathbf{G}_{ZF}(n) = [\mathbf{H}(n)\mathbf{H}^H(n)]^{-1}\mathbf{H}^H(n). \tag{3.18}$$

MMSE Based Equalization

Similarly, the MMSE based equalization scheme outperforms that of ZF in terms of noise power reduction. The MMSE based equalization method is to optimize the MSE as

$$\mathbf{G}_{MMSE}(n) = \arg \min_{\mathbf{G}(n)} \mathrm{E}\{||\hat{\mathbf{s}}(n, i) - \mathbf{s}(n, i)||^2\}, \tag{3.19}$$

where $\mathbf{G}(n)$ is the weighting matrix. Minimizing Eq. (3.19) with respect to $\mathbf{G}(n)$ results in the MMSE equalizer as

$$\mathbf{G}_{MMSE}(n) = \mathbf{H}^H(n)[\mathbf{H}(n)\mathbf{H}^H(n) + \sigma_z^2 \mathbf{I}_M]^{-1}. \tag{3.20}$$

The source symbols have a unit variance and are spatially uncorrelated as $\mathbf{R}_{ss} = \underset{i}{\mathrm{E}}\{\mathbf{s}(n,i)\mathbf{s}^H(n,i)\} = \mathbf{I}_M$, and the noise is spatially uncorrelated with variance σ_z^2 as $\mathbf{R}_{nn} = \underset{i}{\mathrm{E}}\{\mathbf{z}(n,i)\mathbf{z}^H(n,i)\} = \sigma_z^2\mathbf{I}_M$.

V-BLAST Based Equalization

The V-BLAST based equalization [25, 26] method requires a ZF or MMSE equalizer. However, it outperforms both ZF and MMSE in terms of interference cancellation by using ordering to detect the substream with high post-detection SNR. Therefore, the V-BLAST based equalization method can provide a better BER performance than ZF or MMSE based equalization approach. Assuming a number of K transmit antennas, the V-BLAST procedure is summarized below.

For $g = 1 : K$

1. The MMSE or ZF based equalizer is obtained from the channel.
2. The k-th substream ($k = 0, 1, \ldots, K - 1$) in the received signals, corresponding to the k_g-th column of the channel with the highest post-detection SNR (or the k_g-th row of the equalizer with the lowest minimum norm square value), is selected in the g-th detection.
3. The hard estimation of the n_t-th substream is obtained from the equalized signals.
4. The estimated signal on the n_t-th stream is canceled from the received signals.
5. The n_t-th column of the channel or the n_t-th row of the equalizer is set to null.

End.

Equalization and interference cancellation repeat a number of times until all substreams are detected.

ML Based Equalization

The ML based equalization method [23] is powerful, which applies the LS based channel estimate, while searching over the possible transmitted symbols. The ML based equalized signal $\hat{\mathbf{s}}(n, i)$ in the i-th block for MIMO OFDM systems can be written as

$$\hat{\mathbf{s}}(n, i) = \underset{\tilde{\mathbf{s}}(n,i)}{\arg\min} \{\|\mathbf{x}(n, i) - \mathbf{H}(n)\tilde{\mathbf{s}}(n, i)\|_F^2\}, \qquad (3.21)$$

where $\mathbf{x}(n, i)$ and $\tilde{\mathbf{s}}(n, i)$ are the received signal vector and the trial transmitted signal vector, respectively, $\mathbf{H}(n)$ is the channel frequency response matrix on the n-th subcarrier between all K transmit antennas and all M receive antennas. The ML based equalization method requires a number of searches, resulting in very high complexity.

3.2.2 Blind or Semi-Blind Equalization

Unlike training based channel estimation and equalization, no training signal or little information is required in the blind or semi-blind equalization scheme, which can increase spectral efficiency. Basically, blind or semi-blind equalization methods exploit the structure or statistics of the received signals or channel characteristics to recover the transmitted signals [27]. The blind equalization method can be divided into SOS and HOS.

3.2.2.1 SOS Based Blind Equalization

The SOS based blind equalization method is only feasible for non-minimum phase channels which are linear and time-invariant, by using cyclostationary signals with the periodic correlation [28, 29].

The SOS based channel identification method was first proposed in [28] for Single-Input Multiple-Output (SIMO) systems, where the non-minimum phase channel is estimated from the auto-correlation of the received signal.

The noise subspace is another SOS based blind equalization method [30]. The noise and signal subspace can be estimated from the received signal. The subspace based blind channel estimation method was proposed for SIMO systems in [30], and extended to MIMO OFDM systems in [31].

A non-redundant linear precoding was first proposed in [32] and [33], where the specific precoding structure is explored for semi-blind channel estimation at the receiver. In [34], a general precoding scheme was designed for channel estimation, by exploiting the information contained in the signal covariance matrix. It was extended to MIMO systems in [35].

3.2.2.2 HOS Based Blind Equalization

Compared to the SOS, HOS based blind equalization methods are more robust against the Gaussian noise, as the fourth or higher cumulants of the Gaussian noise are equal to zero [36]. Firstly, Constant Modulus Algorithm (CMA) [37] and Finite Alphabet (FA) [38, 39] based equalization methods are described. Then, the focus of the subsection is mainly on the review of ICA and its application to wireless communication systems.

CMA, as one of earliest blind equalization methods, is based on using the magnitude of a signal while ignoring the phase information. It was proposed in [37] for SISO systems with frequency selective fading channels, and extended to MIMO systems with flat fading channels in [40].

The FA based blind equalization method exploits knowledge of the modulation of a signal. This method was employed for a number of wireless communication systems, such as SISO OFDM systems in [38] and MIMO systems with frequency selective fading channels in [39].

Independent Component Analysis

ICA is motivated by the environment in a room where several microphones at different locations receive several people's speaking simultaneously [41]. As an efficient HOS based BSS method, ICA is to use some information in the statistics of the observed signals for estimating the source signals [42].

Let $\mathbf{s} = [s_0, s_1, \ldots, s_{K-1}]^T$ and $\mathbf{y} = [y_0, y_1, \ldots, y_{M-1}]^T$ denote the $K \times 1$ source variables vector and the $M \times 1$ observed variables vector, respectively. Each observed variable is modeled as a linear combination of the source variables. The ICA model can be written as

$$\mathbf{y} = \mathbf{Hs} + \mathbf{z}, \tag{3.22}$$

where $\mathbf{H}$ denotes the $M \times K$ mixing coefficient matrix from all K source variables to all M observed variables, and $\mathbf{z}$ is the noise vector. Without knowledge of the mixing coefficient matrix, independent source variables can be estimated by ICA.

Whitening

In order to use ICA, the received components are required to be as statistically independent as possible. However, the transmitted components are not usually completely uncorrelated. Thus, it is necessary at the receiver to transform the received components into uncorrelated variables. This process is called whitening. It is a useful and simple preprocessing step prior to the use of ICA [42].

Whitening can be performed in connection with Principal Component Analysis (PCA). First, the auto-correlation of the observed variables is derived. Then, a whitening matrix can be obtained from the Eigenvalue Decomposition (EVD) of the auto-correlation. Assuming that the observed variables have a zero mean, i.e., $E\{\mathbf{x}\} = 0$. Let $\mathbf{R}_{xx} = E\{\mathbf{xx}^H\}$ denote the auto-correlation of the observed variables. The EVD of the auto-correlation $\mathbf{R}_{xx}$ can be also expressed as [41]

$$\mathbf{R}_{xx} = \mathbf{U}\Lambda\mathbf{U}^H, \tag{3.23}$$

where the columns of $\mathbf{U}$ include the eigenvectors of $\mathbf{R}_{xx}$, and $\Lambda = \text{diag}\{[\lambda_0, \lambda_1, \ldots \lambda_{M-1}]\}$ is a diagonal matrix, with λ_m denoting the associated eigenvalues.

The $K \times M$ whitening matrix $\mathbf{W}$ can be obtained via the EVD of the auto-correlation $\mathbf{R}_{xx}$, given as

$$\mathbf{W} = \Lambda^{-1/2}\mathbf{U}^H. \tag{3.24}$$

The whitening matrix is designed such that

$$\mathbf{W}E\{\mathbf{xx}^H\}\mathbf{W}^H = \mathbf{I}_K \tag{3.25}$$

The noise compensated whitening matrix $\mathbf{W}_z$ can be given as

$$\mathbf{W}_z = (\boldsymbol{\Lambda} - \sigma_z^2 \mathbf{I}_K)^{-1/2} \mathbf{U}^H, \tag{3.26}$$

with σ_z denoting the noise variance.

HOS-JADE

The objective of whitening is the decorrelation of the received variables. However, whitening is not sufficient to successfully separate the received components. The desired independent components can be obtained by the HOS based ICA [41].

Several known algorithms have been proposed for the ICA. The Bell-Sejnowski algorithm uses the ML based approach [43]. The FastICA method [44] maximizes the non-Gaussianity of the received components and the Joint Approximate Diagonalization of Eigenmatrices (JADE) algorithm [45] employs the higher order based decorrelation.

Among these approaches, the JADE algorithm has been used in wireless communications, and is employed in this thesis, as it requires shorter data sequences than other ICA methods. JADE, as a well established batch algorithm, is based on joint diagonalization of the cumulant matrices of the received components. The 4-th order statistics are used in the JADE algorithm, resulting in the 4-th order cross-cumulant. By minimizing the 4-th order cross-cumulant, the desired independent components can be estimated from the whitened variables [46].

By imposing the higher order decorrelation matrix $\mathbf{V}$ on the whitened signals, the estimate $\hat{\mathbf{s}} = [\hat{s}_0, \hat{s}_1, \ldots, \hat{s}_{K-1}]^T$ can be given as

$$\hat{\mathbf{s}} = \mathbf{V}^H \mathbf{W} \mathbf{x}. \tag{3.27}$$

The unitary matrix $\mathbf{V}$ is the minimization of the 4-th order cross-cumulant, given as

$$\mathbf{V} = \arg\min_{\mathbf{V}} \sum_{jkl=0}^{K-1} |\mathrm{cum}[\hat{s}_j, \hat{s}_j^*, \hat{s}_k, \hat{s}_l^*]|^2, \tag{3.28}$$

where $\mathrm{cum}[\hat{s}_j, \hat{s}_j^*, \hat{s}_k, \hat{s}_l^*]$ is the 4-th order cross-cumulant, given by [41]

$$\begin{aligned}
\mathrm{cum}[\hat{s}_j, \hat{s}_j^*, \hat{s}_k, \hat{s}_l^*] =& \mathrm{E}\{\hat{s}_j\, \hat{s}_j^*\, \hat{s}_k\, \hat{s}_l^*\} - \mathrm{E}\{\hat{s}_j\, \hat{s}_j^*\}\mathrm{E}\{\hat{s}_k\, \hat{s}_l^*\} \\
& - \mathrm{E}\{\hat{s}_j\, \hat{s}_k\}\mathrm{E}\{\hat{s}_j^*\, \hat{s}_l^*\} - \mathrm{E}\{\hat{s}_j\, \hat{s}_l^*\}\mathrm{E}\{\hat{s}_j^*\, \hat{s}_k\}.
\end{aligned} \tag{3.29}$$

Restrictions on Using ICA

ICA is not perfect, as there are a number of assumptions and restrictions [46].

1. Source components are statistically independent.
   That means there is no relationship between source components. This is the principle of applying ICA successfully. Although whitening can achieve the decorrelation of the received components, it could not be used effectively, if a relation exists.

2. Source components have the non-Gaussian distribution.

   The HOS based ICA is only applicable for the case of dealing with the non-Gaussian distribution, as HOS based Gaussian variables vanish [41]. Thus, ICA is essentially applicable for observed variables with the non-Gaussian distribution.

3. The unknown mixing coefficient matrix is invertible.

   In other words, the number of independent source components is equal to or less than the number of observed components.

With respect to the Gaussian or non-Gaussian distribution, there are two scenarios. If source components are Gaussian variables, the distribution of observed components at the receiver is identical to that of source components. In such a case, ICA is not suitable. However, in the case when a mixture of Gaussian and non-Gaussian components exists, only all the non-Gaussian components can be estimated by ICA, while HOS based Gaussian components vanish.

Ambiguity

Similar to other blind methods, ICA gives rise to indeterminacy in the equalized signals in terms of permutation and phase ambiguities. The permutation ambiguity implies the order of the ICA equalized signals is different from that of source components, while the phase ambiguity is the scaling and phase shifting. The relationship between the ICA equalized signals vector $\hat{\mathbf{s}}$ and the source signals vector $\mathbf{s}$ is written as

$$\hat{\mathbf{s}} = \mathbf{PGs}, \tag{3.30}$$

where $\mathbf{P}$ and $\mathbf{G}$ denote the permutation ambiguity and the phase ambiguity, respectively. Unless additional processes are used, the indeterminacy could not be resolved by ICA [41].

Application of ICA to Wireless Communication Systems

In a wireless communication environment, the channel between transmitter and receiver is normally either frequency flat fading or frequency selective fading. For a frequency flat fading channel, ICA can be directly employed on the received signals, as the received signals are a linear combination of the transmitted signals. For frequency selective fading, the channel is termed to be convolutive. In this case, ICA could not be directly applied. However, by using the CP based OFDM technology, the frequency selective fading channel is divided into a number of flat fading channels. In other words, the received siganls in OFDM systems can be transformed to a number of linear instantaneous mixtures. The OFDM based receiver on the n-th subcarrier in the frequency domain is expressed as

$$\mathbf{y}(n, i) = \mathbf{H}(n)\mathbf{s}(n, i) + \mathbf{z}(n). \tag{3.31}$$

The above equation can be viewed as a general OFDM based system model, which is identical to the linear instantaneous mixture ICA model in Eq. (3.22). Therefore, ICA can be directly applied to the received signals $\mathbf{y}(n, i)$, to separate the mixture and recover the transmitted signals on each subcarrier. Spectral efficiency is improved in the ICA based OFDM system, as knowledge of the CSI is not required any more.

As described previously, the drawback of the ICA model results in a possibly different order and phase in the ICA equalized signals on each subcarrier. Further processing is required to resolve the problem. In [47], ICA was used on all subcarriers. The correlation between adjacent subcarriers is used to resolve the frequency dependent permutation problem. However, this approach brings some bit errors and ambiguity error propagation across subcarriers, which could not be solved. In order to avoid this error propagation, a blind receiver structure was proposed in [21], where the correlation between the data on the reference subcarrier and other subcarriers is explored to allow more robust reordering and scaling at the receiver. However, there is a significant performance gap between the higher order modulation scheme and the case with perfect CSI. In [48], a precoding was employed at the transmitter, by superimposing the reference data on to the source data. At the receiver, the permutation and phase ambiguities is eliminated, by using the correlation between reference symbols, without consuming extra bandwidth. However, the precoding based ambiguity elimination method is very sensitive to the precoding constant and the frame length, and has a good BER performance only when the data frame size is large. Also, this method has high computational complexity, particularly for a large number of transmit antennas.

In this chapter, ICA is applied to a number of OFDM based wireless communication systems. Next, ICA is employed in semi-blind MIMO OFDM systems. A number of reference data sequences are carefully designed, and superimposed into the source data sequences, introducing no training overhead and no additional power consumption. This precoding process has two purposes: CFO estimation and ambiguity elimination in the ICA equalized signals. In the following part, ICA is applied to semi-blind CoMP OFDM systems. A short pilot scheme is carefully designed for each user for two purposes: multi-CFO separation and estimation, and ambiguity elimination in the ICA equalized signals. Also, ICA can be employed in semi-blind CA based CoMP OFDMA systems to perform a joint scheme with ICI mitigation and equalization.

References

1. M. Morelli and U. Mengali. Carrier-frequency estimation for transmissions over selective channels. *IEEE Transactions on Communications*, 48(9):1580–1589, Sep. 2000.
2. Y. Tsai, H. Huang, Y. Chen, and K. Yang. Simultaneous multiple carrier frequency offsets estimation for coordinated multi-point transmission in OFDM systems. *IEEE Transaction on Wireless Communications*, 12(9):4558–4568, Sep. 2013.

3. Y. Wu, J. W. M. Bergmans, and S. Attallah. Carrier frequency offset estimation for multiuser MIMO OFDM uplink using CAZAC sequences: performance and sequence optimization. *EURASIP Journal on Wireless Communication and Networking*, 570680-1/11, 2011.

4. Y. Tsai, H. Huang, Y. Chen, and K. Yang. Simultaneous carrier frequency offset estimation for multi-point transmission in OFDM systems. In *Proc. IEEE Global Telecommuniation Conference (Globecom)*, Huston, USA, Dec. 2011.

5. L. Weng, E. K. S. Au, P. W. C. Chen, R. D. Murch, R. S. Cheng, W. H. Mow, and V. K. N. Lau. Effect of carrier frequency offset on channel estimation for SISO/MIMO-OFDM systems. *IEEE Transactions on Wireless Communications*, 6(5):1854–1863, May 2007.

6. P. H. Moose. A technique for orthogonal frequency division multiplexing frequency offset correction. *IEEE Transactions on Communications*, 42(10):2908–2914, Oct. 1994.

7. Y. H. Kim and J. H. Lee. Joint maximum likelihood estimation of carrier and sampling frequency offsets for OFDM systems. *IEEE Transaction on Broadcasting*, 57(2):277–283, Jun. 2011.

8. H. N. Le, T. L. Ngoc, and C. C. Ko. RLS-based joint estimation and tracking of channel response, sampling, and carrier frequency offsets for OFDM. *IEEE Transaction on Broadcasting*, 55(1):84–94, Mar. 2009.

9. R. Frank, S. Zadoff, and R. Heimiller. Phase shift pulse codes with good periodic correlation properties (corresp.). *IRE Transactions on Information Theory*, 8(6):381–382, 1962.

10. D. C. Chu. Polyphase codes with good periodic correlation properties. *IEEE Transactions on Information Theory*, 16(4):531–532, Jul. 1972.

11. N. Sueshiro and M. Hatori. Modulatable orthogonal sequences and their application to SSMA systems. *IEEE Transactions on Information Theory*, 34(1):93–100, Jan. 1988.

12. T. Roman, S. Visuri, and V. Koivunen. Blind frequency synchronization in OFDM via diagonality criterion. *IEEE Transaction on Signal Processing*, 54(8):3125–3135, Aug. 2006.

13. F. Daffara and O. Adami. A new frequency detector for orthogonal multi-carrier transmission techniques. In *Proc. IEEE Vehicular Technology Conference (VTC-95)*, Chicago, USA, Jul. 1995.

14. B. Park, H. Cheon, E. Ko, C. Kang, and D. Hong. A blind OFDM synchronization algorithm based on cyclic correlation. *IEEE Signal Processing Letters*, 11(2):83–85, Feb. 2004.

15. M. Movahhedian, Y. Ma, and R. Tafazolli. Blind CFO estimation for linearly precoded OFDMA uplink. *IEEE Transactions on Signal Processing*, 58(9):4698–4710, Sep. 2010.

16. R. Lin and A. Petropulu. Linear precoding assisted blind channel estimation for OFDM systems. *IEEE Transaction on Vehicular Technology*, 54(3):983–995, May 2005.

17. Y. Zeng, A. R. Leyman, and T. S. Ng. Joint semiblind frequency offset and channel estimation for multiuser MIMO-OFDM uplink. *IEEE Transaction on Communications*, 55(12): 2270–2278, Dec. 2007.

18. P. Sun and L. Zhang. Low complexity pilot aided frequency synchronization for OFDMA uplink transmission. *IEEE Transaction on Wireless Communications*, 8(7):3758–3769, Jul. 2009.

19. Y. S. Cho, J. Kim, W. Y. Yang, and C. G. Kang. *MIMO-OFDM Wireless Communications with MATLAB*. Wiley, Singapore, 2010.

20. S. Coleri, M. Ergen, A. Puri, and A. Bahai. Channel estimation techniques based on pilot arrangement in OFDM systems. *IEEE Transaction on Broadcasting*, 48(3):223–229, Sep. 2002.

21. L. Sarperi, X. Zhu, and A. K. Nandi. Blind OFDM receivers based on independent component analysis for multiple-input multiple-output systems. *IEEE Transactions on Wireless Communications*, 6(11):4079–4089, Nov. 2007.

22. Myeongchoel Shin, Hakju Lee, and Chungyong Lee. Enhanced channel-estimation technique for MIMO-OFDM systems. *IEEE Transaction on Vehicular Technology*, 53(1):261–265, Jan. 2004.

23. S. M. Kay. *Fundamentals of Statistical Signal Processing: Estimation Theory*. Prentice-Hall, Englewood Cliffs, U.S.A., 1993.

24. O. Edfors et al. OFDM channel estimation by singular value decomposition. *IEEE Transaction on Communications*, 46(7):931–939, Jul. 1998.
25. G. J. Foschini. Layered space-time architecture for wireless communication in a fading environment when using multi-element antennas. *Bell Labs Technical Journal*, 1:41–59, Oct. 1996.
26. P. W. Wolniansky, G. J. Foschini, G. D. Golden, and R. A. Valenzuela. VBLAST: An architecture for realizing very high data rates over the rich-scattering wireless channel. In *Proc. URSI International Symposium on Signals, Systems, and Electronics (ISSSE-98)*, pages 295–300, Pisa, Italy, Sep. 1998.
27. L. Tong and S. Perreau. Multichannel blind identification: From subspace to maximum likelihood methods. *Proceedings of the IEEE*, 86(10):1951–1968, Oct. 1998.
28. L. Tong, G. Xu, and T. Kailath. A new approach to blind identification and equalization of multipath channels. In *Proc. 25th Asilomar Conference on Signals, Systems and Computers*, pages 856–860, Pacific Grove, USA, Nov. 1991.
29. J. G. Proakis. *Digital Communications*. McGraw-Hill, Boston, USA, 2001.
30. E. Moulines, P. Duhamel, J. F. Cardoso, and S. Mayrargue. Subspace methods for the blind identification of multichannel FIR filters. *IEEE Transactions on Signal Processing*, 43(2): 516–525, Feb. 1995.
31. W. Bai, C. He, L. Jiang, and H. Zhu. Blind channel estimation in MIMO-OFDM systems. *IEICE Transactions on Communications*, E85-B(9):1849–1853, Sep. 2002.
32. A Petropulu, R. Zhang, and R. Lin. Blind OFDM channel estimation through simple linear precoding. *IEEE Transaction on Wireless Communications*, 54(3):647–655, Mar. 2004.
33. R. Lin and A. Petropulu. Linear precoding assisted blind channel estimation for OFDM systems. *IEEE Transaction on Vehicular Technology*, 54(3):983–995, May 2005.
34. F. Gao and A. Nallanathan. Blind channel estimation for OFDM systems via a generalized precoding. *IEEE Transaction on Vehicular Technology*, 56(3):1155–1164, Jan. 2007.
35. F. Gao and A. Nallanathan. Blind channel estimation for MIMO OFDM systems via nonredundant linear precoding. *IEEE Transaction on Signal Processing*, 55(2):784–789, Jan. 2007.
36. A. Cichocki and S. Amari. *Adaptive Blind Signal and Image Processing*. John Wiley, Chichester, U.K., 2003.
37. J. Treichler and B. Agee. A new approach to multipath correction of constant modulus signals. *IEEE Transactions on Acoustics, Speech, and Signal Processing*, 31(2):459–472, Apr. 1983.
38. S. Zhou and G. B. Giannakis. Finite-alphabet based channel estimation for OFDM and related multicarrier systems. *IEEE Transactions on Communications*, 49(8):1402–1414, Aug. 2001.
39. L. Rota, V. Zarzoso, and P. Comon. Parallel deflation with alphabet-based criteria for blind source extraction. In *Proc. IEEE Workshop on Statistical Signal Processing (SSP)*, Bordeaux, France, Jul. 2005.
40. L. Castedo, C. J. Escudero, and A. Dapena. A blind signal separation method for multiuser communications. *IEEE Transactions on Signal Processing*, 45(5):1343–1348, May 1997.
41. J. Karhunen A. Hyvarinen and E. Oja. *Independent Component Analysis*. John Wiley & Sons, New York, USA, May 2002.
42. A. K. Nandi. *Blind estimation using higher-order statistics*. Kluwer Academic Publishers, Amsterdam, Netherlands, 1999.
43. A. Bell and J. Sejnowski. An information-maximization approach to blind separation and blind deconvolution. *Neural Computation*, 7(6):1129–1159, Feb. 1995.
44. A. Hyvarinen. Fast and robust fixed-point algorithms for independent component analysis. *IEEE Transactions on Neural Networks*, 10(3):626–634, May 1999.
45. J. F. Cardoso. High-order contrasts for independent component analysis. *Neural Computation*, 11(1):157–192, Jan. 1999.
46. J. F. Cardoso. Blind signal separation: Statistical principles. *Proceedings of the IEEE*, 86(10):2009–2025, Oct. 1998.

47. D. Obradovic, N. Madhu, A. Szabo, and C. S. Wong. Independent component analysis for semiblind signal separation in MIMO mobile frequency selective communication channels. In *Proc. IEEE International Conference on Neural Networks (ICNN)*, pages 53–58, Budapest, Hungary, Jul. 2004.
48. J. Gao, X. Zhu, and A. K. Nandi. Non-redundant precoding and PAPR reduction in MIMO OFDM systems with ICA based blind equalization. *IEEE Transactions on Wireless Communications*, 8(6):3038–3049, Jun. 2009.

Chapter 4
Semi-Blind CFO Estimation and Equalization for Single-User MIMO OFDM Systems

4.1 Introduction

In order to enable the successful signal recovery at the receiver for MIMO OFDM systems, there are some challenging issues, such as time synchronization, frequency synchronization, equalization and so on. One of main drawbacks of MIMO OFDM systems is their sensitivity to the CFO due to the mismatch of LOs between transmitter and receiver, which destroys the orthogonality between subcarriers and leads to the ICI. Since multiple transmit or receive antennas can share one LO, there is a single CFO between transmitter and receiver for MIMO OFDM systems.

Most existing work on CFO estimation is assisted by a number of pilot symbols transmitted during the period of preamble. In [1], a CFO estimation approach was proposed using the difference of the phase shifts between two successive identical blocks. In [2], a CFO estimation scheme was proposed, by using the MPP in the time domain. Then, a general periodic pilot in [3] was designed to reduce the computational complexity of the MPP based CFO estimation method. In [4], CAZAC sequences were used for a sub-optimal CFO estimation method. However, the aforementioned work requires a large number of pilots in order to achieve an accurate estimate of the CFO, which reduces spectral efficiency.

Some pieces of work are based on using blind CFO estimation to improve spectral efficiency. In [5], a kurtosis-type criterion based blind CFO estimation method was proposed, based on the measure of the non-Gaussianity of the fourth-order received signal moment. However, its performance depends on the frequency selectivity of a channel. In [6], a blind CFO estimation method was performed by minimizing off-diagonal elements of the auto-covariance matrix of the received signals. However, this method requires a large number of OFDM blocks in a frame to obtain a good MSE performance for CFO estimation, which is not practical in real wireless communication.

© The Author(s) 2015
Y. Jiang et al., *Semi-Blind Carrier Frequency Offset Estimation and Channel Equalization*, SpringerBriefs in Electrical and Computer Engineering, DOI 10.1007/978-3-319-24984-1_4

In this chapter, a semi-blind MIMO OFDM system is presented, with a precoding aided CFO estimation approach and an ICA based equalization structure, where no addition to the total transmit power and no real-time spectral overhead are introduced. A number of reference data sequences are carefully designed offline and are superimposed on to the source data sequences via a non-redundant linear precoding process, killing two birds with one stone. First, assuming that there are K transmit antennas, a total number of K reference data sequences are selected from a pool of carefully designed orthogonal sequences. The search for an optimal CFO is based on minimizing the sum cross-correlations between the received signals with the trial CFO compensation and the rest orthogonal sequences in the pool. Second, with the same reference data sequences, the permutation and quadrant ambiguities in the ICA equalized signals can be eliminated by maximizing the cross-correlation between the ICA equalized signals and the reference signals. The optimal elimination method can be proposed by minimizing the cross-correlation between the different reference data sequences of K transmit antennas. Simulation results show that the proposed semi-blind MIMO OFDM system, with a precoding aided CFO estimation method and an ICA based equalization scheme, achieves a BER performance close to the ideal case with perfect CSI and no CFO. Also, without any pilot introduced, the precoding aided semi-blind CFO estimation method outperforms the CAZAC sequences based CFO estimation approach [4], in terms of BER and MSE performances.

The system model is presented in Sect. 4.2 and the precoding design is presented in Sect. 4.3. The precoding based CFO estimation approach is proposed in Sect. 4.4. The precoding aided ICA based equalization scheme is described in Sect. 4.5. Complexity analysis and simulation results are presented in Sects. 4.6 and 4.7, respectively. Section 4.8 provides a summary.

4.2 System Model

An MIMO OFDM spatial multiplexing system is considered with N subcarriers, K transmit antennas and M receive antennas. The channel is assumed to be block fading and the CSI remains constant for a frame duration of N_s OFDM blocks with a channel length L between the m-th receive antenna and the k-th transmit antenna. The channel frequency response matrix is denoted as $\mathbf{H}_{m,k} = \text{diag}\{[H_{m,k}(0), H_{m,k}(1), \ldots, H_{m,k}(N-1)]\}$, with $H_{m,k}(n)$ being the channel frequency response element on the n-th subcarrier between the k-th transmit antenna and the m-th receive antenna. Each OFDM block is prepended with a CP of length L_{cp} before transmission, and is removed at the receiver to avoid the IBI. Define ϕ as the CFO between the receiver and transmitter, normalized by the sampling frequency. Correspondingly, the diagonal CFO matrix can be denoted as $\boldsymbol{\Phi}^{(\phi)} = \text{diag}\{[1, e^{j2\pi\phi/N}, \ldots, e^{j2\pi(N-1)\phi/N}]\}$.

Let $s_k(n, i)$ denote the symbol on the n-th ($n = 0, 1, \ldots, N-1$) subcarrier in the i-th ($i = 0, 1, \ldots, N_s-1$) OFDM block transmitted by the k-th ($k = 0, 1, \ldots, K-1$)

transmit antenna. Define $\mathbf{s}_k(i) = [s_k(0, i), s_k(1, i), \ldots, s_k(N - 1, i)]^T$ as the signal vector in the i-th OFDM block, transmitted by the k-th transmit antenna. The received signal vector $\mathbf{y}_m(i) = [\mathbf{y}_m(0, i), \mathbf{y}_m(1, i), \ldots, \mathbf{y}_m(N - 1, i)]^T$ in the i-th OFDM block at the m-th receive antenna in the frequency domain after the CP removal, including the frequency offset, is written as

$$\mathbf{y}_m(i) = \varphi^{(\phi)}(i) \sum_{k=0}^{K-1} \mathbf{F} \boldsymbol{\Phi}^{(\phi)} \mathbf{F}^H \mathbf{H}_{m,k} \mathbf{s}_k(i) + \mathbf{z}_m(i), \tag{4.1}$$

where $\varphi^{(\phi)}(i) = e^{j2\pi i\phi N_g/N}$ is the common phase error in the i-th OFDM block, with $N_g = N + L_{cp}$ denoting the total OFDM block length, $\mathbf{F}$ is the $(N \times N)$ DFT matrix with entry (u, v) given by $\mathbf{F}(u, v) = 1/\sqrt{N}e^{-j2\pi uv/N}, (u, v = 0, 1, \ldots, N - 1)$, and $\mathbf{z}_m(i)$ is the AWGN vector whose entries are i.i.d. complex Gaussian random variables with a zero mean and a variance.

According to Eq. (4.1), the CFO of ϕ leads to a common phase error in different OFDM blocks, and destroys the orthogonality between subcarriers. Also, the received signals $\mathbf{y}_m(i)$ at the m-th receive antenna are combined with K transmitted signals causing the ISI. Some methods were proposed to estimate the CFO and channel in the literature [2–4]. However, these approaches require a large number of training sequences to obtain accurate CFO and channel estimation.

4.3 Precoding Design

A number of reference data sequences are designed via a precoding process at the transmit side for two purposes: (1) CFO estimation, and (2) ambiguity elimination in the ICA equalized signals.

4.3.1 Precoding

The source data sequences are superimposed by a number of reference data sequences via a non-redundant precoding process. The reference data vector of length N for the k-th transmit antenna in the i-th OFDM block can be expressed as $\mathbf{d}_{\text{ref},k}(i) = [d_{\text{ref},k}(0, i), d_{\text{ref},k}(1, i), \ldots, d_{\text{ref},k}(N - 1, i)]^T$. Accordingly, the source data vector is written as $\mathbf{d}_k(i) = [d_k(0, i), d_k(1, i), \ldots, d_k(N - 1, i)]^T$. Both the source data $\mathbf{d}_k(i)$ and the reference data $\mathbf{d}_{\text{ref},k}(i)$ are assumed to be of unit average power. They are also assumed to be uncorrelated with each other. The entries of $\mathbf{d}_{\text{ref},k}(i)$ are i.i.d., and have the same discrete probability distribution as the entries of $\mathbf{d}_k(i)$. The resulting transmit signal vector $\mathbf{s}_k(i)$, is given by

$$s_k(i) = \frac{1}{\sqrt{1+a^2}}[d_k(i) + ad_{\text{ref},k}(i)], \tag{4.2}$$

where a ($0 \leq a \leq 1$) is the precoding constant, to trade off power allocation between the source data $d_k(i)$ and the reference data $d_{\text{ref},k}(i)$. A scaling factor $\frac{1}{\sqrt{1+a^2}}$ is applied to keep the transmit power unchanged.

4.3.2 Reference Data Design

The reference data sequences shown in Eq. (4.2) play a key role in CFO estimation and ambiguity elimination in the ICA equalized signals, requiring no extra total transmit power and no extra bandwidth consumption. Let $D_{\text{ref},k} = [d_{\text{ref},k}(0), d_{\text{ref},k}(1),$
$\ldots, d_{\text{ref},k}(N_s - 1)]$ denote the $N \times N_s$ reference matrix for the k-th transmit antenna. Define Q as the $U \times N_s$ projection operator matrix. On the one hand, the precoding based CFO estimation depends on minimizing the sum cross-correlations between the reference data sequences $D_{\text{ref},k}$ and the projection operator Q. On the other hand, minimizing the cross-correlation between different reference data sequences of K transmit antennas provides an optimal ambiguity elimination in the ICA equalized signals, to maximize the cross-correlation between the reference signals and the ICA equalized signals. Therefore, the optimal reference data design is to minimize the cost function as follows

$$J = \frac{1}{(N^2 + NU)(K-1)K} \sum_{k=0}^{K-1} \sum_{l=0,l\neq k}^{K-1} \left(||D_{\text{ref},k}Q^H||_F^2 + ||D_{\text{ref},k}D_{\text{ref},l}^H||_F^2 \right). \tag{4.3}$$

The optimal solution that minimizes Eq. (4.3) is a set of reference data sequences that are orthogonal to each other and orthogonal to the projection operator Q, so that $J = 0$ as $D_{\text{ref},k}Q^H = 0$ and $D_{\text{ref},k}D_{\text{ref},l}^H = 0$ ($k \neq l$).

The $N_s \times N_s$ Hadamard matrix M_{N_s} [7, 8] can provide a solution to Eq. (4.3), where any two different rows are orthogonal to each other, i.e., $M_{N_s}M_{N_s}^H = N_s I_{N_s}$. Without loss of generality, a number of K different rows of M_{N_s} are randomly selected as the reference data sequences for K transmit antennas. For the reference data sequences design of user k, the selected row in M_{N_s} is placed in each row of $D_{\text{ref},k}$, as shown below

$$D_{\text{ref},k} = [\underbrace{M_{N_s}(i_k,:)^T, M_{N_s}(i_k,:)^T, \ldots, M_{N_s}(i_k,:)^T}_{N \text{ subcarriers}}]^T, \tag{4.4}$$

where $M_{N_s}(i_k,:)$ ($0 \leq i_k \leq (N_s - 1)$) denotes the i_k-th row of M_{N_s}, allocated to each row of $D_{\text{ref},k}$. Then, the remaining U rows are to form the projection operator matrix Q, as shown below

$$\mathbf{Q} = \underbrace{[\mathbf{M}_{N_s}(i_q^{(0)}, :)^T, \mathbf{M}_{N_s}(i_q^{(1)}, :)^T, \dots, \mathbf{M}_{N_s}(i_q^{(U-1)}, :)^T]^T}_{U \text{ rows}}, \tag{4.5}$$

where $\mathbf{M}_{N_s}(i_q^{(u)}, :)$ $(0 \leqslant i_q^{(u)} \leqslant (N_s - 1); u = 0, 1, \dots, U - 1; i_q^{(u)} \neq i_k)$ denotes the $i_k^{(u)}$-th row of $\mathbf{M}_{N_s}$, allocated to the u-th row of $\mathbf{Q}$. Therefore, according to the reference data sequences design in Eq. (4.4) and the projection operator design in Eq. (4.5), the correlation between the reference data sequences and the projection operator, and the correlation between the reference data sequences of antennas k and l are respectively given by

$$\mathbf{D}_{\text{ref},k}\mathbf{Q}^H = \mathbf{0}_{N \times U}. \tag{4.6}$$

$$\mathbf{D}_{\text{ref},k}\mathbf{D}_{\text{ref},l}^H = \begin{cases} N\mathbf{I}_N & k = l \\ \mathbf{0}_{N \times N} & k \neq l \end{cases}. \tag{4.7}$$

First, according to Eqs. (4.4), (4.5), and (4.6), the different reference data sequences of K transmit antennas are orthogonal to the projection operator, i.e., $\mathbf{D}_{\text{ref},k}\mathbf{Q}^H = \mathbf{0}$. This allows CFO estimation to be performed at the receiver. Second, according to Eqs. (4.4) and (4.7), the reference data sequences of different transmit antennas are orthogonal to each other, i.e., $\mathbf{D}_{\text{ref},k}\mathbf{D}_{\text{ref},l}^H = \mathbf{0}$ as $k \neq l$. This provides an optimal ambiguity elimination in the ICA equalized signals.

4.3.3 Discussion of Precoding Constant

The precoding constant a in Eq. (4.2) varies from 0 to 1, and offers a trade-off of power allocation between the source data and the reference data. The variations of the precoding constant have impact on MSE and BER performances. On the one hand, in order to provide a good MSE performance for CFO estimation, a larger valued precoding constant is preferred, which will increase the power of the reference data, i.e. $|a| \to 1$. However, in such a case, the transmit power of the source data decreases, resulting in a lower SNR. Thus, a large precoding constant causes a BER degradation. On the other hand, to achieve a good BER performance, a smaller value of precoding constant is preferred. However, if the precoding constant is too small, i.e. $|a| \to 0$, neither CFO estimation nor ambiguity elimination in the ICA equalized signals can be performed satisfactorily, as the power for the reference data could not be preserved. Therefore, the precoding constant must be carefully selected to balance the power between the source data and the reference data, while providing good MSE and BER performances.

4.4   Precoding Based Semi-Blind CFO Estimation

Using the features of the reference data sequences and the projection operator in Sect. 4.3, a semi-blind precoding aided CFO estimation method is proposed for MIMO OFDM systems, where a number of carefully designed reference data sequences are superimposed to the source data without changing the total transmit power. The precoding based CFO estimation method is performed by exploring the cross-correlation between the trial CFO compensated received signals and the projection operator. As the reference data sequences and the projection operator can be designed offline and known in advance at the receiver, the precoding based CFO estimation method introduces no additional training overhead, and thus has high bandwidth efficiency.

The effect of CFO in this chapter gives rise to two issues: an ICI and a common phase error. Unlike traditional searching based CFO estimation methods [6, 9], which require the ICI and common phase error compensation with the trial CFO, the proposed semi-blind CFO estimation method only requires the common phase error compensation. Therefore, large savings in computational complexity can be made, as the complex trial ICI compensation is not required for each search.

With a CFO value ϕ, the signal vector $\check{\mathbf{s}}_k^{(\phi)}(i)$ with the common phase error in the i-th OFDM block is written as

$$\check{\mathbf{s}}_k^{(\phi)}(i) = \varphi^{(\phi)}(i)\mathbf{s}_k(i). \tag{4.8}$$

Given a trial value $\tilde{\phi}$ of the CFO, the common phase error compensation may be performed in the frequency domain at the receiver. Using Eqs. (4.1) and (4.8), the received signal vector $\tilde{\mathbf{y}}_m(i)$ with the common phase error compensation in the i-th block is expressed as

$$\tilde{\mathbf{y}}_m(i) = \sum_{k=0}^{K-1} \mathbf{F}\boldsymbol{\Phi}^{(\phi)}\mathbf{F}^H\mathbf{H}_{m,k}\check{\mathbf{s}}_k^{(\phi-\tilde{\phi})}(i) + \tilde{\mathbf{z}}_m(i), \tag{4.9}$$

where $\tilde{\mathbf{z}}_m(i) = \varphi^{(-\tilde{\phi})}(i)\mathbf{z}_m(i)$ is the noise with respect to the common phase error compensation using the trial CFO value.

According to Eqs. (4.2) and (4.8), after collecting N_s OFDM blocks, the k-th transmitted signal matrix $\check{\mathbf{S}}_k^{(\phi-\tilde{\phi})} = [\check{\mathbf{s}}_k^{(\phi-\tilde{\phi})}(0), \ldots, \check{\mathbf{s}}_k^{(\phi-\tilde{\phi})}(N_s - 1)]$ with respect to the common phase error compensation with the trail CFO value can be written as

$$\check{\mathbf{S}}_k^{(\phi-\tilde{\phi})} = \frac{1}{\sqrt{1+a^2}}[\check{\mathbf{D}}_k^{(\phi-\tilde{\phi})} + a\check{\mathbf{D}}_{\text{ref},k}^{(\phi-\tilde{\phi})}], \tag{4.10}$$

where $\check{\mathbf{D}}_k^{(\phi-\tilde{\phi})} = [\check{\mathbf{d}}_k^{(\phi-\tilde{\phi})}(0), \ldots, \check{\mathbf{d}}_k^{(\phi-\tilde{\phi})}(N_s - 1)]$ with vector $\check{\mathbf{d}}_k^{(\phi-\tilde{\phi})}(i)$ in the i-th block written as $\check{\mathbf{d}}_k^{(\phi-\tilde{\phi})}(i) = [\varphi^{(\phi-\tilde{\phi})}(i)d_k(0, i), \ldots, \varphi^{(\phi-\tilde{\phi})}(i)d_k(N - 1, i)]^T$, and

$\check{\mathbf{D}}_{\text{ref},k}^{(\phi-\tilde{\phi})} = [\check{\mathbf{d}}_{\text{ref},k}^{(\phi-\tilde{\phi})}(0), \ldots, \check{\mathbf{d}}_{\text{ref},k}^{(\phi-\tilde{\phi})}(N_s - 1)]$ with vector $\check{\mathbf{d}}_{\text{ref},k}^{(\phi-\tilde{\phi})}(i)$ in the i-th block written as $\check{\mathbf{d}}_{\text{ref},k}^{(\phi-\tilde{\phi})}(i) = [\varphi^{(\phi-\tilde{\phi})}(i)d_{\text{ref},k}(0,i), \ldots, \varphi^{(\phi-\tilde{\phi})}(i)d_{\text{ref},k}(N-1,i)]^T$.

Similarly, after collecting N_s OFDM blocks, the received signals are written in the form as $\tilde{\mathbf{Y}}_m = [\tilde{\mathbf{y}}_m(0), \tilde{\mathbf{y}}_m(1), \ldots, \tilde{\mathbf{y}}_m(N_s - 1)]$, and are given by

$$\tilde{\mathbf{Y}}_m = \sum_{k=0}^{K-1} \mathbf{F}\boldsymbol{\Phi}^{(\phi)}\mathbf{F}^H \mathbf{H}_{m,k}\check{\mathbf{S}}_k^{(\phi-\tilde{\phi})} + \tilde{\mathbf{Z}}_m, \tag{4.11}$$

where $\tilde{\mathbf{Z}}_m = [\tilde{\mathbf{z}}_m(0), \ldots, \tilde{\mathbf{z}}_m(N_s - 1)]$.

Define $\boldsymbol{\Gamma}_m$ as the correlation between the received signals $\tilde{\mathbf{Y}}_m$ with the common phase error compensation at the m-th receive antenna and the projection operator $\mathbf{Q}$. It can be written as

$$\boldsymbol{\Gamma}_m = \tilde{\mathbf{Y}}_m\mathbf{Q}^H. \tag{4.12}$$

Substituting Eqs. (4.11) into (4.12) yields

$$\boldsymbol{\Gamma}_m = \frac{1}{\sqrt{1+a^2}} \sum_{k=0}^{K-1} \Big[\mathbf{F}\boldsymbol{\Phi}^{(\phi)}\mathbf{F}^H\mathbf{H}_{m,k}$$
$$\times (\check{\mathbf{D}}_k^{(\phi-\tilde{\phi})}\mathbf{Q}^H + a\check{\mathbf{D}}_{\text{ref},k}^{(\phi-\tilde{\phi})}\mathbf{Q}^H) \Big] + \tilde{\mathbf{Z}}_m\mathbf{Q}^H. \tag{4.13}$$

The CFO ϕ can be estimated using the correlation in Eq. (4.6). With a correct CFO trial value, i.e., $\phi = \tilde{\phi}$, the correlation between the reference data sequences and the projection operator is zero, i.e., $\check{\mathbf{D}}_{\text{ref},k}\mathbf{Q}^H = \mathbf{0}$. In such a case, the correlation $\boldsymbol{\Gamma}_m$ between the received signals and the projection operator can be written as

$$\boldsymbol{\Gamma}_m = \frac{1}{\sqrt{1+a^2}} \sum_{k=0}^{K-1} [\mathbf{F}\boldsymbol{\Phi}^{(\phi)}\mathbf{F}^H\mathbf{H}_{m,k}\mathbf{D}_k\mathbf{Q}^H] + \tilde{\mathbf{Z}}_m\mathbf{Q}^H. \tag{4.14}$$

Comparing Eqs. (4.13) and (4.14), it can be derived that the correct common phase error compensation may result in a $\boldsymbol{\Gamma}_m$ with the minimum energy. The high-accuracy CFO estimation can be performed by minimizing the sum cross-correlations from all M receive antennas. Note that if the precoding constant a is far too small, the energy of $\boldsymbol{\Gamma}_m$ will not be very different, to the variations of the trial CFO. In order to reduce the effect, the precoding constant a is set to be just small enough. Also, a large size of $\mathbf{Q}$ is required, that means more data sequences are used to form the projection operator. In general, the larger the value of U, the more accurate the CFO estimation. The estimate $\hat{\phi}$ of CFO ϕ can be obtained via a search according to

$$\hat{\phi} = \underset{\tilde{\phi} \in [-0.4,\, 0.4]}{\arg\ \min} \sum_{m=0}^{M-1} ||\boldsymbol{\Gamma}_m||_F^2, \qquad (4.15)$$

where $|| \cdot ||_F^2$ denotes the Frobenius norm. As the proposed precoding based CFO estimation approach only requires knowledge of the reference data sequences at the receiver which can be obtained offline, it provides higher spectral efficiency than other pilots based CFO estimation methods.

4.5 Precoding Aided Semi-Blind ICA Based Equalization

With the CFO estimate value $\hat{\phi}$, the CFO compensated received signals vector $\hat{\mathbf{y}}_m(i) = [\hat{y}_m(0, i), \hat{y}_m(1, i), \dots, \hat{y}_m(N - 1, i)]^T$ in the i-th block becomes

$$\hat{\mathbf{y}}_m(i) = \varphi^{(\phi-\hat{\phi})}(i) \sum_{k=0}^{K-1} \left[\mathbf{F}\boldsymbol{\Phi}^{(\phi-\hat{\phi})}\mathbf{F}^H\mathbf{H}_{m,k}\mathbf{s}_k(i) \right] + \mathbf{F}\boldsymbol{\Phi}^{(-\hat{\phi})}\mathbf{F}^H\mathbf{z}_m(i). \qquad (4.16)$$

Assuming perfect CFO estimation and compensation, the received signals vector $\hat{\mathbf{y}}(n, i) = [\hat{y}_0(n, i), \hat{y}_1(n, i), \dots, \hat{y}_{M-1}(n, i)]^T$ on the n-th subcarrier in the i-th block can be written as

$$\hat{\mathbf{y}}(n, i) = \mathbf{H}(n)\mathbf{s}(n, i) + \hat{\mathbf{z}}(n, i), \qquad (4.17)$$

where $\mathbf{H}(n) = [\mathbf{H}_0^T(n), \mathbf{H}_1^T(n), \dots, \mathbf{H}_{K-1}^T(n)]^T$ with $\mathbf{H}_k(n) = [H_{0,k}, H_{1,k}, \dots, H_{M-1,k}]^T$ is the $M \times K$ channel frequency response matrix between all K transmit antennas and M receive antennas on the n-th subcarrier, $\mathbf{s}(n, i) = [s_0(n, i), s_1(n, i), \dots, s_{K-1}(n, i)]^T$ denotes the K transmit signals vector on the n-th subcarrier in the i-th block, and $\hat{\mathbf{z}}(n, i)$ is the noise vector.

4.5.1 ICA Based Equalization

ICA [10], an efficient HOS based BSS technique, is employed to recover $\mathbf{s}(n, i)$ from Eq. (4.17), by maximizing the non-Gaussianity of the received signals. The JADE algorithm [11] requires shorter data sequences than other ICA numerical algorithms, and is employed in this chapter, since the received signals are a linear mixture of the transmitted signals on each subcarrier as shown in Eq. (4.17). However, the ICA model has a major drawback in terms of permutation and phase ambiguities in the ICA equalized signals, which can be resolved by maximizing the correlation between the reference data and the ICA equalized data. Let $\boldsymbol{\pi}(n) = [\pi_0(n), \pi_1(n), \dots, \pi_{K-1}(n)]$ denote the permuted order in the ICA separated signals on the n-th subcarrier. After using ICA on the received signals, the equalized signals vector $\tilde{\mathbf{s}}(n, i) = [\tilde{s}_{\pi_0(n)}(n, i), \tilde{s}_{\pi_1(n)}(n, i), \ldots, \tilde{s}_{\pi_{K-1}(n)}(n, i)]^T$ is expressed as

$$\tilde{\mathbf{s}}(n, i) = \mathbf{L}(n)\mathbf{P}(n)\mathbf{B}(n)\mathbf{s}(n, i) \qquad (4.18)$$

where $\mathbf{L}(n)$ is the phase deviation matrix, $\mathbf{P}(n)$ is the permutation ambiguity matrix and $\mathbf{B}(n)$ is the phase ambiguity matrix.

4.5.2 Ambiguity Elimination

The phase deviation in $\tilde{\mathbf{s}}(n, i)$ can be resolved by de-rotating the phase of each data stream. The de-rotated signals vector $\check{\mathbf{s}}(n, i) = [\check{s}_{\pi_0(n)}(n, i), \check{s}_{\pi_1(n)}(n, i), \ldots, \check{s}_{\pi_{K-1}(n)}(n, i)]^T$ is written as

$$\check{\mathbf{s}}(n, i) = \mathbf{L}^{-1}(n)\tilde{\mathbf{s}}(n, i), \qquad (4.19)$$

where $\mathbf{L}^{-1}(n) = \mathrm{diag}\{\frac{\alpha_{\pi_0(n)}(n)}{|\alpha_{\pi_0(n)}(n)|}, \frac{\alpha_{\pi_1(n)}(n)}{|\alpha_{\pi_1(n)}(n)|}, \ldots, \frac{\alpha_{\pi_{K-1}(n)}(n)}{|\alpha_{\pi_{K-1}(n)}(n)|}\}$ is the de-rotation matrix, with the de-rotation factor $\alpha_{\pi_k(n)}(n)$ being

$$\alpha_{\pi_k(n)}(n) = \left\{\frac{1}{N_s}\sum_{i=0}^{N_s-1}[\tilde{s}_{\pi_k(n)}(n, i)]^4\right\}^{-\frac{1}{4}} e^{j\frac{\pi}{4}}. \qquad (4.20)$$

This de-rotation factor is obtained from the statistics of the ICA equalized signals with the Quadrature Phase Shift Keying (QPSK) modulation [12, 13].

The cross-correlation $\rho_{\pi_k(n),k}(n)$ between the $\pi_k(n)$-th ICA separated substream and the reference data sequences of the k-th transmit antenna on the n-th subcarrier over a number of N_s blocks is defined as

$$\rho_{\pi_k(n),k}(n) = \frac{1}{N_s}\sum_{i=0}^{N_s-1}\check{s}_{\pi_k(n)}(n, i)d^*_{\mathrm{ref},k}(n, i). \qquad (4.21)$$

According to the reference data sequences design, the cross-correlation between different reference data sequences of K transmit antennas is minimized, to ensure that the different reference data sequences are orthogonal to each other as shown in Eq. (4.7). This provides an optimal solution to find the correct order $\pi_k^{cor}(n)$ by maximizing the correlation $\rho_{\pi_k(n),k}(n)$ between the ICA equalized signals and the reference signals on the n-th subcarrier as

$$\pi_k^{cor}(n) = \arg\max_{\pi_k(n)}|\rho_{\pi_k(n),k}(n)|. \qquad (4.22)$$

Table 4.1 Analytical computational complexity (EQ: equalization)

Item		Order of complexity
Precoding		NKN_s
CFO estimation		$NMUN_s(1/\Delta)$
ICA EQ	ICA(JADE)	$N(M^2 + K^4N_s + K^5)$
	Phase correction	NKN_s
	Ambiguity elimination	$NN_s(K!)$

After ordering, the correlation $\rho_{\pi_k(n),k}(n)$ can be used again to eliminate the phase ambiguity in the equalized signals. The transmitted signals can be estimated by

$$\hat{s}(n, i) = [\mathbf{B}(n)]^{-1}[s_{\pi_0^{cor}(n)}(n, i), s_{\pi_1^{cor}(n)}(n, i), \ldots, s_{\pi_{K-1}^{cor}(n)}(n, i)]^T, \qquad (4.23)$$

where $\mathbf{B}(n)$ is the diagonal phase ambiguity matrix, with entry $b_{\pi_k^{cor}(n)}(n)$ being

$$b_{\pi_k^{cor}(n)} = \left[e^{-j\frac{\pi}{4}} \text{sign}\left(\frac{\rho_{\pi_k^{cor}(n),k}(n)}{|\rho_{\pi_k^{cor}(n),k}(n)|} e^{j\frac{\pi}{4}} \right) \right]^{-1}. \qquad (4.24)$$

4.6   Complexity Analysis

In this section, the computational complexity of the proposed semi-blind precoding aided CFO estimation method and the ICA based equalization scheme is investigated for MIMO OFDM systems, in terms of the number of complex multiplications. Define Δ as the step size between -0.4 and 0.4, resulting in a total number of $\frac{1}{\Delta}$ searches to obtain a reliable CFO estimate. Table 4.1 shows that the complexity of the precoding based CFO estimation depends on the size of the projection operator matrix $\mathbf{Q}$ which consists of U selected data sequences. A larger valued U will result in better CFO estimation accuracy, however, this will increase the computational complexity. Therefore, the value of U should be selected carefully to make a trade-off between performance and complexity. In the ICA based equalization, the computational complexity grows with an increasing number of transmit antennas. The precoding based ambiguity elimination requires a search for all possibilities to find a correct order in the ICA equalized signals.

4.7   Simulation Results

The simulation results are used to demonstrate the performance of the proposed semi-blind precoding aided CFO estimation method and the ICA based equalization structure for MIMO OFDM systems with $N = 64$ subcarriers, $K = 2$ transmit antennas and $M = 2$ receive antennas. A CP of length $L_{cp} = 15$ is used. Clarke's block fading channel model [14] is employed, where the CSI remains constant

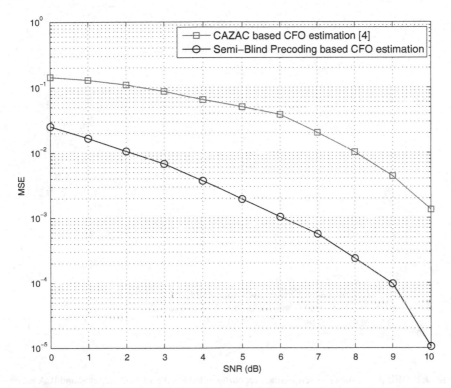

Fig. 4.1 MSE performance of the proposed semi-blind precoding based CFO estimation for MIMO OFDM systems

during a frame. The RMS delay spread is $T_{\text{RMS}} = 1.4$ normalized to the sampling period. The resulting channel impulse response length is $L = 5$. Each data frame consists of $N_s = 256$ OFDM blocks with the QPSK modulation. The precoding constant is set to be $a = 0.25$ as in [13], resulting in a decrease of less than 0.4 dB in the effective SNR. A number of $U = 182$ sequences are selected from the Hadamard matrix to form the projection operator $\mathbf{Q}$. The normalized CFO is randomly generated between -0.4 and 0.4. The step size is set to $\Delta = 0.01$ resulting in a total number of 80 searches.

The MSE between the true and estimated CFOs, is defined as MSE = E $\{(\phi - \hat{\phi})^2\}$, where E denotes expectation. Figure 4.1 shows the MSE performance of the proposed semi-blind precoding based CFO estimation method, compared with that of the CAZAC sequences based CFO estimation approach [4]. There are a total number of 32 blocks for the CAZAC sequences based CFO estimation method. The precoding based CFO estimation method provides a much better MSE performance, around 6 dB gain, compared to the CAZAC sequences based approach, from SNR = 0 dB to SNR = 10 dB. It is observed that the MSE performance achieved by the precoding based CFO estimation approach is almost zero at a SNR higher than 10 dB. This is not the case for the CAZAC sequences based CFO estimation method which can achieve a good performance only at high SNRs.

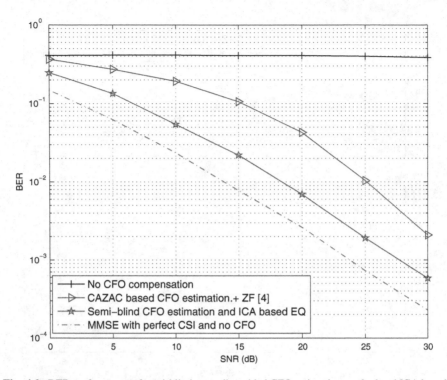

Fig. 4.2 BER performance of semi-blind precoding aided CFO estimation method and ICA based equalization structure for MIMO OFDM systems, in comparison to the MMSE based equalization method with perfect CSI and no CFO (EQ: equalization)

Figure 4.2 shows the BER performance of the proposed semi-blind MIMO OFDM system, in comparison with the MMSE based equalization method with perfect CSI and no CFO at the receiver. Also, the CAZAC sequences based CFO estimation method [4] is used for comparison, with the LS based channel estimation plus the ZF based equalization. There are a number of 32 blocks for the CAZAC sequences based CFO estimation and the LS based channel estimation. The proposed semi-blind CFO estimation method and the ICA based structure can provide a BER performance close to the MMSE based equalization method with perfect CSI and no CFO. It also has a better BER performance than the CAZAC sequences based CFO estimation method. Without compensation, the ZF based equalization has an error floor at high BER across the whole range of SNRs. Compared with the CAZAC sequences based CFO estimation approach, only offline knowledge of the precoding is required in the proposed CFO estimation method, introducing no additional training overhead. Therefore, the proposed precoding based CFO estimation method provides good performance and high bandwidth efficiency.

4.8 Summary

In this chapter, a semi-blind ICA based MIMO OFDM system is proposed, where a number of reference data sequences are superimposed into the source data sequences via a non-redundant linear precoding process at the transmitter, with no additional power consumption and no training overhead used. Only knowledge of the reference data sequences is required at the receiver, and can be obtained offline.

These reference data sequences are carefully designed to kill two birds with one stone: CFO estimation and ambiguity elimination in the ICA equalized signals. A cost function is formed for the reference data sequences and the projection operator. By minimizing the cost function, the optimal reference data sequences and the projection operator can be designed, and obtained from the Hadamard matrix. The precoding constant introduced in the precoding process is discussed, which provides a trade-off of power between the source data and the reference data, introducing no additional transmit power.

The sum cross-correlations between the reference data sequences and the projection operator are used to perform CFO estimation at the receiver. A number of selected sequences are discussed to form the projection operator. In general, the larger the number of selected sequences, the more accurate the CFO estimation. Different from traditional search based CFO estimation methods [6, 9], requiring both the ICI and common phase error compensation for each search within the range of CFO values, the proposed semi-blind precoding based CFO estimation method only requires the common phase error compensation. Therefore, it can provide lower complexity, as the complex trial ICI compensation is not needed.

After CFO estimation and compensation, ICA is employed for equalization at the receiver on each subcarrier. However, there are permutation and phase ambiguities in the ICA equalized signals. According to the reference data sequences design, the cross-correlation between different reference data sequences of transmit antennas is minimized so that the different reference data sequences are orthogonal. This provides an optimal solution to eliminate the permutation ambiguity by maximizing the correlation between the ICA equalized signals and the reference signals. This correlation can be used again on each subcarrier for phase ambiguity elimination.

Regarding simulation results, the proposed semi-blind MIMO OFDM system is a viable alternative to training based systems. The reasons can be given as follows. First, the proposed precoding based CFO estimation approach outperforms the CAZAC sequences based CFO estimation method [4] in terms of MSE and BER performances. Second, the proposed semi-blind MIMO OFDM system, with the precoding aided CFO estimation method and the ICA based equalization scheme, achieves a BER performance close to the ideal case with perfect CSI and no CFO. Also, compared to training based approaches, the proposed semi-blind MIMO OFDM system can provide higher bandwidth efficiency. Therefore, it demonstrates high performance and high spectral efficiency.

However, the precoding based method is proposed for a single CFO only. When there are multiple CFOs, this method is not suitable due to the required complex multi-dimensional search. Also, the precoding based ambiguity elimination method requires a search for all possibilities to find a correct order in the ICA equalized signals. The computational complexity grows rapidly with the increasing number of transmit antennas. In the next chapter, a multi-CFO estimation method is proposed for semi-blind ICA based multiuser CoMP OFDM systems. A short pilot is introduced, dividing a complex multi-dimensional search into a number of one-dimensional searches. The same pilot is used again to eliminate the permutation and phase ambiguities simultaneously, rather than sequentially as in the precoding based method, achieving lower complexity.

References

1. M. Morelli and U. Mengali. Carrier-frequency estimation for transmissions over selective channels. *IEEE Transactions on Communications*, 48(9):1580–1589, Sep. 2000.
2. G. Xing, M. Shen, and H. Liu. Frequency offset and I/Q imbalance compensation for direct-conversion receivers. *IEEE Transactions on Wireless Communications*, 4(2):673–680, Mar. 2005.
3. H. Lin, X. Zhu, and K. Yamashita. Pilot aided low complexity CFO and I/Q imbalance compensation for OFDM systems. In *Proc. IEEE International Conference on Communications (ICC)*, Beijing, China, Jun. 2008.
4. Y. Wu, J. W. M. Bergmans, and S. Attallah. Carrier frequency offset estimation for multiuser MIMO OFDM uplink using CAZAC sequences: performance and sequence optimization. *EURASIP Journal on Wireless Communication and Networking*, 570680-1/11, 2011.
5. Y. Yao and G. B. Giannakis. Blind carrier frequency offset estimation in SISO, MIMO, and multiuser OFDM systems. *IEEE Transaction on Communications*, 53(1):173–183, Jan. 2005.
6. T. Roman, S. Visuri, and V. Koivunen. Blind frequency synchronization in OFDM via diagonality criterion. *IEEE Transaction on Signal Processing*, 54(8):3125–3135, Aug. 2006.
7. Q. K. Trinh and P. Z. Fan. Construction of multilevel Hadamard matrices with small alphabet. *Electronic Letter*, 44(21):1250–1252, Oct. 2008.
8. A. C. Lossifides. Complex orthogonal coded binary transmission with amicable Hadamard matrices over rayleigh fading channels. In *Proc. IEEE Symposium on Computers and Communications*, pages 335–340, Kerkyra, Greece, Jun. 2011.
9. P. Sun and L. Zhang. Low complexity pilot aided frequency synchronization for OFDMA uplink transmission. *IEEE Transaction on Wireless Communications*, 8(7):3758–3769, Jul. 2009.
10. J. Karhunen A. Hyvarinen and E. Oja. *Independent Component Analysis*. John Wiley & Sons, New York, USA, May 2002.
11. J. F. Cardoso. High-order contrasts for independent component analysis. *Neural Computation*, 11(1):157–192, Jan. 1999.
12. L. Sarperi, X. Zhu, and A. K. Nandi. Blind OFDM receivers based on independent component analysis for multiple-input multiple-output systems. *IEEE Transactions on Wireless Communications*, 6(11):4079–4089, Nov. 2007.
13. J. Gao, X. Zhu, and A. K. Nandi. Non-redundant precoding and PAPR reduction in MIMO OFDM systems with ICA based blind equalization. *IEEE Transactions on Wireless Communications*, 8(6):3038–3049, Jun. 2009.
14. A. Goldsmith. *Wireless Communications*. Cambridge University Press, London, U.K., 2005.

Chapter 5
Semi-Blind Multi-CFO Estimation and Equalization for Multiuser CoMP OFDM Systems

5.1 Introduction

Recently, the CoMP structure [1], in which geographically separated BSs are connected to each other to jointly process the signals of multiple users at cell edge, has been recommended for implementation of the LTE-Advanced system [1]. The CoMP structure can reduce interference and enhance system throughput.

As described before, in CoMP systems, there are multiple CFOs between BSs and users, which can degrade BER performance, unless there is high-accuracy multi-CFO estimation and compensation. The precoding based method proposed in the previous chapter is used for single CFO estimation only. It is not suitable for cases with multiple CFOs, since a complex multi-dimensional search is required. The high computational complexity means that the method becomes practically impossible.

In the literature, there are a number of CFO estimation approaches. In [2], an ML based CFO estimation method was proposed. However, its computational complexity is still high for CoMP systems with multiple CFOs. In [3], Moose used periodic training sequences for CFO estimation, which however is not applicable to CoMP systems with multiple CFOs. In [4], a precoding aided multi-CFO estimation approach was proposed for uplink OFDMA systems. However, it is not applicable to OFDM based CoMP systems, where multiple users share subcarriers and cause the interference to each other. So far, little work in the literature has considered multi-CFO estimation for CoMP systems, where frequency synchronization is required between all BSs and users, resulting in challenging multi-CFO estimation. In [5] and [6], an optimal ML based multi-CFO estimation approach was proposed for downlink CoMP systems. However, it suffers an error floor at high SNRs due to the MUI. In [7], a suboptimal CFO estimation method based on using CAZAC sequences was proposed for multiuser MIMO OFDM systems. However, it is a biased approach, and requires long CAZAC sequences for accurate CFO estimation.

© The Author(s) 2015
Y. Jiang et al., *Semi-Blind Carrier Frequency Offset Estimation and Channel Equalization*, SpringerBriefs in Electrical and Computer Engineering,
DOI 10.1007/978-3-319-24984-1_5

Furthermore, the aforementioned approaches use different pilots for CFO estimation and channel estimation.

Multi-cell channel estimation and equalization are important and challenging in CoMP systems [8]. ICA [9], an HOS based BSS approach, has been proven effective for blind or semi-blind channel equalization [10, 11]. The main drawback of ICA is the ambiguity (such as phase and permutation ambiguities) in the equalizer output signals. In [10], the correlations between a reference subcarrier and all other subcarriers were used to resolve the ICA ambiguity. However, it can introduce error propagation across subcarriers. In [11], a precoding was used, where the reference data is superimposed in the source data to allow ambiguity elimination. However, it must be operated on a relatively large data frame to achieve a good performance.

In this chapter, a low-complexity semi-blind multiuser CoMP OFDM system is proposed in the uplink with multi-CFO estimation and ICA based equalization, where a short pilot for each user is designed carefully for two purposes. On the one hand, using the pilot structure, a complex multi-dimensional search for multiple CFOs is divided into a number of low-complexity mono-dimensional searches. On the other hand, the cross-correlation between the transmitted and the received pilots is explored to allow the simultaneous elimination of permutation ambiguity and quadrant ambiguity in the ICA equalized signals.

- First, a set of generic Semi-Orthogonal Pilot (SOP) is proposed for multiple users, which enables separation of multiple CFOs at each BS. Thus, a complex multi-dimension search for multiple CFOs [2] is divided into a number of low-complexity mono-dimensional searches at each BS. Multi-CFO estimation is then performed independently at different BSs based on correlations of the pilots on selected subcarriers. As these CFO estimates are only required at a reference BS for the calculation of carrier frequency adjustments, the CoMP system burden is low.
- Second, besides multi-CFO estimation, the proposed SOP are also used for the elimination of permutation and quadrant ambiguities induced by the ICA based equalization, exploring the correlation between the original and equalized pilots. This is less complex than the precoding aided method in [11], as the permutation and quadrant ambiguities can be eliminated simultaneously, rather than sequentially.
- Third, the proposed SOP based approach has a high degree of freedom and allows a trade-off between complexity, performance and spectral efficiency. They can also be easily extended to the MIMO CoMP scenario.
- Simulation results show that, the proposed semi-blind CoMP system with multiple CFOs provides a BER performance close to the ideal case with perfect CSI and no CFO. The SOP based CFO estimation method significantly outperforms Moose's approach [3] and the CAZAC sequences based approach [7], in terms of BER and MSE of CFO estimation performances.

The system model is presented in Sect. 5.2 and the pilot design is presented in Sect. 5.3. The proposed multi-CFO estimation approach and the ICA based equalization scheme are described in Sects. 5.4 and 5.5, respectively. Complexity analysis and simulation results are presented in Sects. 5.6 and 5.7, respectively. Section 5.8 provides a summary of this chapter.

5.2 System Model

An uplink CoMP OFDM system is considered, where K users simultaneously transmit independent QPSK symbols of unit energy to M BSs, by using N subcarriers in each OFDM block, as illustrated in Fig. 5.1. Energy per bit is denoted by E_b. It is assumed that each user and BS are equipped with a single antenna. Define $\mathbf{s}_k(i) = [s_k(0,i),\ldots,s_k(N-1,i)]^T$ as the signal vector of user k $(k = 0,\ldots,K-1)$ in block i $(i = 0,\ldots,N_s-1)$, with $s_k(n,i)$ denoting the symbol on subcarrier n $(n = 0,\ldots,N-1)$. We define f_{B_m} and f_{U_k} as the carrier frequencies at BS m $(m = 0,\ldots,M-1)$ and user k, respectively. Let $\phi_{B_m,U_k} = (f_{B_m} - f_{U_k})/\Delta f$ denote the normalized CFO caused by the mismatch of LOs between BS m and user k, where Δf is the subcarrier spacing. Similarly, define ϕ_{B_m,B_t} $(m,t = 0,\ldots,M-1)$ and ϕ_{U_k,U_l} $(k,l = 0,\ldots,K-1)$ as the CFOs between BSs m and t, and between users k and l, respectively. The CFO-induced ICI matrix between BS m and user k is defined as $\mathbf{C}_{m,k} = \sum_{v=0}^{N-1}(e^{j2\pi v\phi_{B_m,U_k}}\mathbf{F}_v)$, where $\mathbf{F}_v$ is an $N \times N$ matrix with entry (a,b) given by $\mathbf{F}_v(a,b) = (1/\sqrt{N})e^{-j2\pi(a-b)/N}$, $(a,b = 0,\ldots,N-1)$. The channel

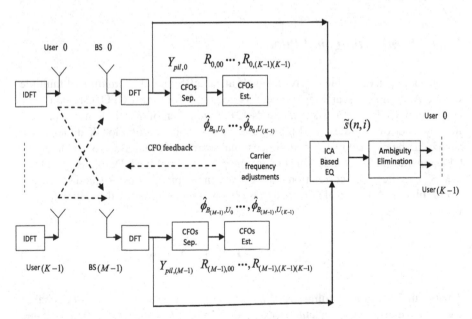

Fig. 5.1 Block diagram of the multiuser CoMP OFDM system with semi-blind multi-CFO estimation and ICA based equalization (DFT: discrete Fourier transform, IDFT: inverse discrete Fourier transform, Sep.: separation, Est.: estimation, EQ: equalization)

is assumed to remain constant for a frame duration of N_s OFDM blocks, with a channel length L. Let $\mathbf{H}_{m,k}$ denote the $N \times N$ diagonal channel frequency response matrix between BS m and user k. To avoid the IBI, each OFDM block is prepended with a CP of length L_{cp} ($L_{cp} \geqslant L$). The received frequency-domain signal vector $\mathbf{y}_m(i)$ in block i at BS m can be written as [12]

$$\mathbf{y}_m(i) = \sum_{k=0}^{K-1} \mathbf{C}_{m,k} \mathbf{H}_{m,k} \mathbf{s}_k(i) + \mathbf{z}_m(i), \tag{5.1}$$

where $\mathbf{z}_m(i)$ is the AWGN vector whose entries are i.i.d. complex Gaussian random variables with a zero mean and a variance of N_0.

5.3   Pilot Design

A pilot of P OFDM blocks is designed for each user. Let $\mathbf{S}_{\text{pil},k} = [\mathbf{s}_k(0), \ldots, \mathbf{s}_k (P-1)]$ denote the k-th user's pilot of size $N \times P$. In the initial stage, the set of K pilots are transmitted simultaneously for multi-CFO estimation and carrier frequency adjustments. In the data transmission stage, the same pilots take the first P blocks of a data frame of N_s blocks, and are used to eliminate ambiguity in the ICA equalized signals.

5.3.1   Semi-orthogonal Pilot

As shown in Eq. (5.1), the received signals at each BS include a mixture of multiple CFOs. A complex searching process is required for multi-CFO estimation by conventional CFO estimation methods [2]. A number of K CFOs are separated first by a set of well designed pilots at each BS. Thus, a complex K-dimensional search can be divided into K one-dimensional searches at each BS. To enable multi-CFO separation, the correlation between the pilots is employed. Define $\mathbf{R}_{kl} = (1/P) \mathbf{S}_{\text{pil},k} \mathbf{S}_{\text{pil},l}^H$ as the $N \times N$ correlation matrix between the pilots of users k and l averaged over P blocks. An optimal pilot design is obtained by minimizing the following cost function

$$J = \frac{1}{K^2 N^2} \sum_{k=0}^{K-1} \left(||\mathbf{R}_{kk} - \mathbf{I}_N||_2^F + \sum_{l=0,l\neq k}^{K-1} ||\mathbf{R}_{kl}||_2^F \right). \tag{5.2}$$

The optimal solution to minimize Eq. (5.2) is a set of pilots that are orthogonal to each other in the frequency and user domains, so that $J = 0$ as $\mathbf{R}_{kl} = \mathbf{0}$ ($k \neq l$) and $\mathbf{R}_{kk} = \mathbf{I}_N$. However, the minimum required length of the Orthogonal Pilot (OP) is large when the numbers of subcarriers and users are large.

To allow a good trade-off between spectral efficiency, CFO estimation accuracy and performance of ICA ambiguity elimination, a number of pilots are designed to be orthogonal in the user domain, and semi-orthogonal in the frequency domain. Therefore, they are referred to as SOP.

The features of a $P \times P$ Hadamard matrix $\mathbf{M}_P$ [13, 14] are utilized in the pilot design, where any two different rows of $\mathbf{M}_P$ are orthogonal to each other, i.e., $\mathbf{M}_P\mathbf{M}_P^H = P\mathbf{I}_P$. Without loss of generality, for the pilot design of user k, $(Q + 1)$ different rows of $\mathbf{M}_P$ are randomly selected and placed in the first $(Q + 1)$ rows of $\mathbf{S}_{\mathrm{pil},k}$, where $0 \leqslant Q \leqslant \min(P, N) - 1$. For the purpose of SOP design, the last $(N - Q - 1)$ rows of $\mathbf{S}_{\mathrm{pil},k}$ are allowed to be the repetitions of row Q, as illustrated below

$$\mathbf{S}_{\mathrm{pil},k} = [\underbrace{\mathbf{M}_P(i_k^{(0)}, :)^T, \ldots, \mathbf{M}_P(i_k^{(Q-1)}, :)^T}_{Q \text{ subcarriers}}, \underbrace{\mathbf{M}_P(i_k^{(Q)}, :)^T, \ldots, \mathbf{M}_P(i_k^{(Q)}, :)^T}_{(N-Q) \text{ subcarriers}}]^T, \quad (5.3)$$

where $\mathbf{M}_P(i_k^{(q)}, :)$ $(0 \leqslant i_k^{(q)} \leqslant (P - 1), q = 0, 1, \ldots, Q)$ denotes the $i_k^{(q)}$-th row of $\mathbf{M}_P$, allocated to the q-th row of $\mathbf{S}_{\mathrm{pil},k}$. Ensuring $(Q + 1)K$ different rows of $\mathbf{M}_P$ are selected for the pilots of K users, the correlation matrix between the pilots of users k and l averaged over P blocks is given by

$$\mathbf{R}_{kl} = \begin{cases} \begin{bmatrix} \mathbf{I}_Q & \mathbf{0}_{Q \times (N-Q)} \\ \mathbf{0}_{(N-Q) \times Q} & \mathbf{1}_{(N-Q)} \end{bmatrix} & k = l \\ \mathbf{0}_{N \times N} & k \neq l \end{cases} \quad (5.4)$$

According to Eq. (5.4), different users' pilots are orthogonal to each other, i.e., $\mathbf{R}_{kl} = \mathbf{0}$ $(k \neq l)$. This enables simultaneous multi-CFO separations at all BSs. The pilot symbols within a few blocks of a user are orthogonal to each other on part of subcarriers, indicated by the $(2N - Q - 1)Q$ zero elements in $\mathbf{R}_{kl}$ when $k = l$. Substituting Eqs. (5.4) into (5.2) yields $J = (N - Q)(N - Q - 1)/N^2$. When $Q = N - 1$, the SOP becomes the OP, and $\mathbf{R}_{kk}$ becomes an identity matrix $\mathbf{I}_N$.

5.3.2 Requirement on Pilot Length

In the SOP design, $(Q + 1)K$ different rows out of P rows of the Hadamard matrix $\mathbf{M}_P$ are selected for K users. Thus, the pilot length P must satisfy $P \geqslant (Q + 1)K$. As the size of the Hadamard matrix must be multiples of 4 [13, 14], the condition can be satisfied as $P = 4\eta$ $(\eta = 1, 2, \ldots)$. The minimum value of Q is 1 when $K = 2\eta$. With $\eta = 1$, the minimum required pilot length is $P = 2K$, twice of the number of users. In the special case of OP, i.e., $Q = N - 1$, the minimum required pilot length is $P = KN$. Thus, the SOP achieves a $[N/(Q + 1)]$-fold reduction in training overhead over the OP. This is equal to $N/2$ when the minimum value of $Q = 1$ is used. It will be shown in simulation results that the SOP provides a BER performance close to that of the OP.

5.4   SOP Based Multi-CFO Estimation

Using the features of the SOP presented in the previous section, K CFOs are separated first, and then are estimated independently at each BS. Thus, a complex K-dimensional search is divided into K one-dimensional searches at each BS. The proposed CFO estimation is based on the correlation of each user's pilot in the frequency domain. The CFO estimates only need to be sent to a reference BS for the calculation of how much a carrier frequency should be adjusted at each user and each of the rest of BSs. Thus, the system burden is low. Note that the total number of CFOs depends on the numbers of users and BSs only. In a MIMO CoMP system, where there are K users with multiple transmit antennas each and M BSs with multiple receive antennas each, there are still a total number of MK CFOs. Therefore, the proposed approach can be easily extended to the MIMO scenario.

5.4.1   Multi-CFO Separation

Let $\mathbf{Y}_{\mathrm{pil},m} = [\mathbf{y}_m(0),\ldots,\mathbf{y}_m(P-1)]$ denote the $N \times P$ received pilot at BS m, which contains the mixing effects of multiple CFOs and channels associated with K users. As the pilots are orthogonal in the user domain, to separate K CFOs, each user's pilot is used as a projection operator. Define $\mathbf{R}_{m,kk} = (1/P)\mathbf{Y}_{\mathrm{pil},m}\mathbf{S}_{\mathrm{pil},k}^H$ as the $N \times N$ correlation matrix between the received mixture of pilots from K users at BS m and the transmit pilot of user k averaged over P blocks. It includes the combined effects of CFO and channel frequency response between user k and BS m. Using Eqs. (5.1) and (5.4), $\mathbf{R}_{m,kk}$ can be written as

$$\mathbf{R}_{m,kk} = \mathbf{C}_{m,k}\mathbf{H}_{m,k}\mathbf{R}_{kk} + \tilde{\mathbf{Z}}_{\mathrm{pil},m}, \qquad (5.5)$$

where $\tilde{\mathbf{Z}}_{\mathrm{pil},m} = (1/P)\mathbf{Z}_{\mathrm{pil},m}\mathbf{S}_{\mathrm{pil},k}^H$ is the $N \times N$ noise matrix averaged over P blocks, with $\mathbf{Z}_{\mathrm{pil},m} = [\mathbf{z}_m(0),\ldots,\mathbf{z}_m(P-1)]$. It can be observed from Eq. (5.5) that multiple CFOs are successfully separated using the user-domain orthogonality of the SOP. Thus, there is no MUI in the separated CFOs.

5.4.2   Multi-CFO Estimation

Define $\tilde{\phi}_{\mathrm{B}_m,\mathrm{U}_k}$ as a trial value of $\phi_{\mathrm{B}_m,\mathrm{U}_k}$ and $\tilde{\mathbf{C}}_{m,k} = \sum_{v=0}^{N-1}(e^{j2\pi v\tilde{\phi}_{\mathrm{B}m,\mathrm{U}_k}}\mathbf{F}_v)$ as the corresponding ICI matrix applied to $\mathbf{R}_{m,kk}$. Let $\tilde{\mathbf{R}}_{m,kk} = \tilde{\mathbf{C}}_{m,k}\mathbf{R}_{m,kk}$. Using Eq. (5.5), it can be written as

$$\tilde{\mathbf{R}}_{m,kk} = \tilde{\mathbf{C}}_{m,k}^H\mathbf{C}_{m,k}\mathbf{H}_{m,k}\mathbf{R}_{kk} + \tilde{\mathbf{C}}_{m,k}^H\tilde{\mathbf{Z}}_{\mathrm{pil},m}, \qquad (5.6)$$

The CFO ϕ_{B_m,U_k} can be estimated using the frequency-domain semi-orthogonality of user k's pilot. The pairs of orthogonal subcarriers are indicated by $(2N - Q - 1)Q$ zero-valued off-diagonal elements in $\mathbf{R}_{kk}$ of Eq. (5.4). With a perfect CFO trial value and no noise, the corresponding off-diagonal elements in $\tilde{\mathbf{R}}_{m,kk}$ are also zero. Note that the channel matrix $\mathbf{H}_{m,k}$ is diagonal and therefore does not affect the off-diagonal elements in $\tilde{\mathbf{R}}_{m,kk}$. Thus, the proposed CFO estimation is performed by minimizing the sum power of selected off-diagonal elements in $\tilde{\mathbf{R}}_{m,kk}$. To allow a trade-off between complexity and performance, only the correlations between the first D $(0 < D \leqslant N)$ subcarriers are considered. The search area is limited to the first D rows and D columns of $\tilde{\mathbf{R}}_{m,kk}$, excluding any elements corresponding to the nonzero elements in $\mathbf{R}_{kk}$ of Eq. (5.4). Thus, only $(2D-Q-1)Q$ out of $(2N-Q-1)Q$ qualified correlation coefficients in $\tilde{\mathbf{R}}_{m,kk}$ need to be considered. As the normalized CFO value is in the range of $[-0.5, \ 0.5)$, a search needs to be performed over the range to obtain the CFO estimate $\hat{\phi}_{B_m,U_k}$ according to

$$\hat{\phi}_{B_m,U_k} = \arg \min_{\tilde{\phi}_{B_m,U_k}} \left|\left|\tilde{\mathbf{R}}_{m,kk}(1:D,1:D) \odot [\mathbf{1}_D - \mathbf{R}_{kk}(1:D,1:D)]\right|\right|_F^2, \qquad (5.7)$$

where $\odot$ and $||\cdot||_F^2$ denote the Hadamard product and Frobenius norm, respectively. With a step size of Δ, a total of $(1/\Delta)$ searches are needed.

5.4.3 Carrier Frequency Adjustments

After CFO estimation, the carrier frequencies are adjusted at BSs and user ends. Due to the mixture of multiple users' CFOs, the carrier frequencies of BSs could not be adjusted independently. A feedback correction method is used, similar to that in [15]. Taking the carrier frequency of BS t as a reference, the carrier frequencies of all K users and the rest of $(M-1)$ BSs are adjusted to the reference frequency. To achieve this, the KM CFO estimates $\hat{\phi}_{B_m,U_k}$ obtained in the previous section are sent to BS t $(t = 0,\ldots,M-1)$ for the calculation of the carrier frequency difference between each user or BS and the reference BS. Those CFO values are then fed back to individual users and BSs for frequency adjustments.

Frequency adjustment at BS m: Direct frequency synchronization between two BSs of a CoMP system requires an additional receiver to be installed at a BS to listen to the other's transmission [15]. The frequency difference between BSs m and t $(m \neq t)$ is calculated by taking the averaged difference between each user's two CFO estimate values at BSs m and t

$$\bar{\phi}_{B_m,B_t} = \frac{1}{K} \sum_{k=0}^{K-1} (\hat{\phi}_{B_m,U_k} - \hat{\phi}_{B_t,U_k}). \qquad (5.8)$$

This method does not require extra hardware installation.

Frequency adjustment at user k: Define $\bar{\phi}_{U_k,U_l} = (1/M) \sum_{m=0}^{M-1}(\hat{\phi}_{B_m,U_l} - \hat{\phi}_{B_m,U_k})$ as the estimated carrier frequency difference between users k and l, using the CFO estimates for users k and l obtained at all BSs. The carrier frequency difference $\bar{\phi}_{U_k,B_t}$ between user k and BS t is calculated by taking the difference between $\bar{\phi}_{U_k,U_l}$ and $\hat{\phi}_{B_t,U_l}$ averaged over all possible values of l $(l \neq k)$:

$$\bar{\phi}_{U_k,B_t} = \frac{1}{K-1} \sum_{l=0, l \neq k}^{K-1} (\bar{\phi}_{U_k,U_l} - \hat{\phi}_{B_t,U_l}). \tag{5.9}$$

5.5   ICA Based Equalization

After CFO estimation and carrier frequency adjustments, the received signals are a linear mixture of the non-Gaussian transmit signals. Thus, ICA [9] can be employed for semi-blind equalization. The SOP are used to resolve the permutation and phase ambiguities in ICA equalized signals.

The ICA algorithm JADE [16] can be applied on a frame of N_s OFDM blocks to separate the received signals on each subcarrier. Thus, the system model is expressed in terms of each subcarrier. Define $s(n, i) = [s_0(n, i), \ldots, s_{K-1}(n, i)]^T$ as the transmit signal vector on subcarrier n in block i. The ICA equalized signal vector $\tilde{s}(n, i) = [\tilde{s}_0(n, i), \ldots, \tilde{s}_{K-1}(n, i)]^T$ is expressed as [11]

$$\tilde{s}(n, i) = P(n)G(n)s(n, i), \tag{5.10}$$

where $P(n)$ and $G(n)$ denote the $K \times K$ matrices accounting for the permutation ambiguity and phase ambiguity, respectively. Define $\pi_k(n)$ as the permuted order of user k on subcarrier n $(\pi_k(n), k = 0 \ldots, K-1)$. The $\pi_k(n)$-th row of the permutation ambiguity $P(n)$ is expressed as

$$[P(n)](\pi_k(n), :) = [\mathbf{0}_{1 \times k} \ 1 \ \mathbf{0}_{1 \times (K-k-1)}]. \tag{5.11}$$

The phase ambiguity is expressed as

$$G(n) = \text{diag}\{[g_0(n), \ldots, g_{K-1}(n)]\}, \tag{5.12}$$

with $g_k(n)$ denoting the phase shift for user k on subcarrier n.

5.5.1   Phase Correction

The phase ambiguity in the ICA equalized signal can be resolved by de-rotating the phase of the data frame from user k by

$$\check{s}_k(n, i) = g_k(n)^{-1} \tilde{s}_k(n, i), \tag{5.13}$$

where the phase shift $g_k(n)$ for user k on subcarrier n is written as

$$g_k(n) = \frac{|\alpha_k(n)|}{\alpha_k(n)}, \tag{5.14}$$

with $\alpha_k(n)$ denoting the factor obtained from the statistical characteristics of $\tilde{s}_k(n, i)$ for the QPSK modulation, given by [11]

$$\alpha_k(n) = \{\frac{1}{N_s} \sum_{i=0}^{N_s-1} [\tilde{s}_k(n, i)]^4\}^{-\frac{1}{4}} e^{j\frac{\pi}{4}}. \tag{5.15}$$

5.5.2 Permutation and Quadrant Ambiguities Elimination

Although phase correction in Eq. (5.14) resolves the phase ambiguity in the equalized signal $\tilde{s}_k(n, i)$, it introduces a phase rotation of $\theta = \frac{\pi}{2}l, l \in \{0, 1, 2, 3\}$ in $\check{s}_k(n, i)$, which is referred to as quadrant ambiguity. This can be resolved together with permutation ambiguity, by using the correlation between the equalized and original pilots, since the ICA equalized source symbols have the same permutation and phase ambiguities as the pilot symbols. Let $\rho_{\pi_k(n),k}(n)$ denote the cross-correlation between the $\pi_k(n)$-th ICA separated pilot $\check{s}_{\pi_k(n)}(n, i)$ and the k-th user's pilot $s_k(n, i)$ on subcarrier n averaged over P blocks, with the trial quadrant ambiguity $\tilde{\theta}_{\pi_k(n)}$. It is expressed as

$$\rho_{\pi_k(n),k}(n) = \frac{1}{P} \sum_{i=0}^{P-1} \{[\check{s}_{\pi_k(n)}(n, i)e^{j\tilde{\theta}_{\pi_k(n)}}]s_k^*(n, i)\}. \tag{5.16}$$

Theorem 1. *If the real part of $\rho_{\pi_k(n),k}(n)$ is maximized, the permutation and phase ambiguities correction has the highest probability of being correct. $\rho_{\pi_k(n),k}(n) = 1$ if there is no error (proof: see Appendix A). Therefore, the correct user order $\pi_k^{cor}(n)$ and the correct phase rotation $\theta_{\pi_k^{cor}(n)}$ to eliminate the quadrant ambiguity, can be found as*

$$[\pi_k^{cor}(n), \theta_{\pi_k^{cor}(n)}] = \underset{\pi_k(n), \tilde{\theta}_{\pi_k(n)}}{\arg \max} \Re\{\rho_{\pi_k(n),k}(n)\}. \tag{5.17}$$

$\Re\{\cdot\}$ *denotes the real part of a complex number. The permutation and quadrant ambiguities can be eliminated simultaneously, rather than sequentially as in [11]. The reason can be shown in Appendix A. The SOP based approach also has a faster convergence speed than the approach as in [11], as demonstrated in the simulation results.*

Table 5.1 Analytical computational complexity (P: Pilot length, N: Number of subcarriers, K: Number of users, M: Number of BSs, Δ: Step size for CFO search, Q: Number of subcarriers for SOP design, D: Number of subcarriers considered for CFO search)

Item			Order of complexity
Multi-CFO estimation	CFOs separation		$(2P-1)N^2KM$
	CFOs estimation	$D \leqslant Q+1$	$(1/\Delta)(2DN-1)DKM$
		$D > Q+1$	$(2/\Delta)(2D-Q-1)QKMN$
	Carrier frequency adjustments		$2K+M-1$
ICA equalization	ICA (JADE) [9]		$(K^5 + K^4N_s + M^2)N$
	Phase correction		KNN_s
	Permutation and quadrant ambiguity elimination		$4KNP$

5.6 Complexity Analysis

In Table 5.1, the computational complexity of the proposed semi-blind system for a frame of N_s OFDM blocks is presented, in terms of the number of complex additions and multiplications. The complexity of the SOP based multi-CFO estimation method increases linearly with the numbers of BSs and users, rather than exponentially as in [2]. Also, the proposed multi-CFO estimation approach has lower complexity than [7], where a complex multi-dimensional search is required for suitable CAZAC sequences. Compared to the special case of OP with $Q = N$, $P = KN$ and $D = N$, the SOP based multi-CFO separation and estimation with $Q = 1, P = 2K$ and $D = N/2$ requires approximately $0.6N^2$ times less complexity. This is around 2400 times less when $N = 64$, which is the same used in the simulation setup in the next section. Compared with the precoding based method in [11], the SOP based ambiguity elimination method requires $N_s(K-1)!/(4P)$ times less complexity. This is because the SOP based ambiguity elimination method requires the assistance of only a small number of pilot blocks, rather than all the received blocks in each data frame as in [11].

5.7 Simulation Results

In this section, simulation results are used to demonstrate the performance of the proposed SOP based multi-CFO estimation and the ICA based equalization scheme for CoMP systems, with $K = 2$ users and $M = 2$ BSs. It is assumed that a data frame contains $N_s = 256$ OFDM blocks of $N = 64$ QPSK symbols, with a data rate of 32 Mbps. A CP of length $L_{cp} = 16$ is used. Clarke's block fading channel model [17] is used, where the channel follows an exponential delay profile with a root mean square delay spread of 330 ns and a channel length of $L = 5$. With the SOP based design, the first $P = 4$ blocks (minimum required) in a frame are used as pilot, resulting in a training overhead of 1.6 %. While for OP, a special case of SOP, a minimum pilot length of $P = 128$ is required, resulting in a training overhead of 50 %. A set of $Q = 1$ and $D = 32$ is used for SOP, while $Q = 64$ and $D = 64$ for OP. A step size of $\Delta = 0.01$ is used to search for each CFO estimate within the range of $[-0.5, 0.5)$.

For performance comparison, two other CFO estimation approaches are used: Moose's method [3] and the CAZAC sequences based method [7]. They are combined with LS channel estimation [12] and ZF equalization. Using the SOP based CFO estimation, the performance of the ICA based equalization is also compared with that of LS channel estimation [12] and ZF equalization. The performance of the ICA based equalization with the SOP aided ambiguity elimination is also compared to that of [11], using a precoding constant value of 0.25. Furthermore, ZF and MMSE based equalization schemes with perfect CSI and no CFO are used as benchmarks.

In Fig. 5.2, the BER performance of the proposed SOP and ICA based scheme is demonstrated, in comparison with the other approaches mentioned above. The CFO values are set to be 0.3 for the links between user 1 and both BSs, and −0.3

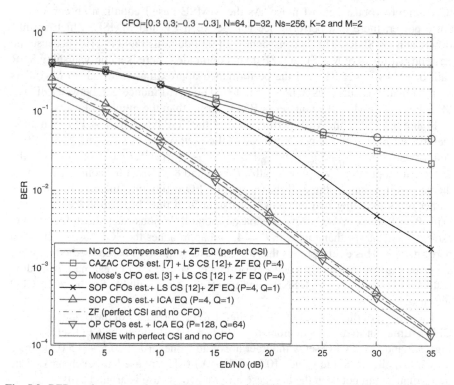

Fig. 5.2 BER performance of the SOP based multi-CFO estimation method and ICA based equalization structure, in comparison to the ZF and MMSE based equalization with perfect CSI and no CFO, with $K = 2$ users and $M = 2$ BSs (est.: estimation, CE: channel estimation, EQ: equalization)

for the links between user 2 and both BSs. The CFO set is written in the form of [0.3 0.3; −0.3 −0.3]. Without CFO compensation, ZF equalization suffers an error floor at a BER = 0.4, even in the case of perfect CSI. Using the same pilot length ($P = 4$) and the same LS and equalization (ZF) methods, the SOP based multi-CFO estimation approach significantly outperforms Moose's [3] and the CAZAC sequences based [7] CFO estimation methods. Using the SOP based CFO estimation, the ICA based equalization demonstrates an SNR gain of around 11 dB at a BER = 2×10^{-3} over LS channel estimation + ZF equalization. This verifies that the ICA based equalization is more robust against noise. With a training overhead of 1.6 %, the SOP based multi-CFO estimation and ICA based equalization scheme can provide a BER performance close to the ideal case with perfect CSI and no CFO. The performance of the SOP based scheme is also close to that of the OP based scheme, with a 32-fold reduction in training overhead, and a 2400-fold reduction in complexity of multi-CFO separation and estimation. The OP based multi-CFO estimation and ICA based equalization scheme even slightly outperforms ZF equalization with perfect CSI and no CFO. This is because ZF equalization may enhance the noise power, while ICA is an HOS based method and the fourth or higher-order cumulants of Gaussian noise are equal to zero [18]. Thus, ICA is more robust against noise. As the MMSE based equalization can reduce noise power, there is a performance gap between ZF and MMSE, which shrinks slowly at high SNRs. This is consistent with that demonstrated in [19], because the error probabilities of MMSE and ZF do not coincide. There is a non-vanishing SNR gain of MMSE over ZF, and the performance approximation of ZF to MMSE is asymptotically accurate as SNR → ∞ [19]. The performance gap between ZF and MMSE reduces when the number of BSs is larger than the number of users.

Figure 5.3 shows the MSE performance of the SOP based multi-CFO estimation method, compared to the performances of the methods in [7] and [3]. The MSE is defined as MSE = $\frac{1}{MK} \sum_{m=0}^{M-1} \sum_{k=0}^{K-1} E\{|\hat{\phi}_{B_m,U_k} - \phi_{B_m,U_k}|^2\}$, where E{·} denotes the expectation operator. Two pilot lengths, $P = 4$ and 8 are used for comparison. With the same pilot length, the SOP based multi-CFO estimation method significantly outperforms the other two methods, consistent with the observation from Fig. 5.2. With $P = 8$, the CAZAC sequences based method demonstrates much better CFO estimation accuracy than that with $P = 4$, and approaches the SOP based method in performance. This demonstrates that the CAZAC sequences based method requires a longer pilot to achieve a good performance. Moose's CFO estimation method suffers an error floor at high SNRs, and therefore is not suitable for the multi-CFO scenario.

The impact of different CFO sets on the MSE performance of the proposed multi-CFO estimation method is demonstrated in Fig. 5.4. Besides the CFO set used for Figs. 5.2 and 5.3, a CFO set with larger absolute values, [0.5 −0.5; −0.5 0.5], and a set with smaller absolute values, [0.1 −0.1; −0.1 0.1], are used as benchmarks. The performance gap between the best case and the worst case is around 8 dB.

Figure 5.5 shows the impact of data frame length on the BER performance of the ICA based semi-blind equalization scheme, in the absence of CFOs, with $E_b/N_0 = 10$ and 20 dB, respectively. The SOP aided method and the precoding aided method [11] reach the steady state with approximately $N_s = 48$ and 200

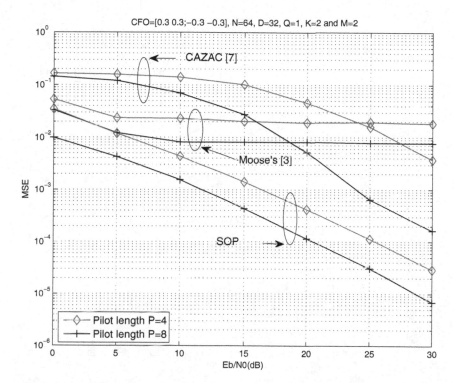

Fig. 5.3 MSE performance of the SOP based multi-CFO estimation method for multiuser CoMP OFDM system, in comparison to CAZAC based and Moorse's CFO estimation approaches, with $K = 2$ users and $M = 2$ BSs

blocks per frame, respectively. This is because the ambiguity elimination in [11] depends on the reference data superimposed in the source data of a whole frame. Thus, the longer the frame length, the better the ambiguity elimination performance, while the SOP aided ambiguity elimination is performed using a small number of pilot blocks. Thus, the proposed ambiguity elimination performance is more robust against frame length change and channel variations. The SOP aided method also has a better steady-state performance than the precoding aided method, because some transmit power is allocated to the reference data from the source data in the precoding process at the transmitter.

5.8 Summary

In this chapter, a semi-blind ICA based multi-user CoMP OFDM system is proposed, where a short pilot is attached with the source data at the transmitter for the simultaneous transmission, introducing a low training overhead.

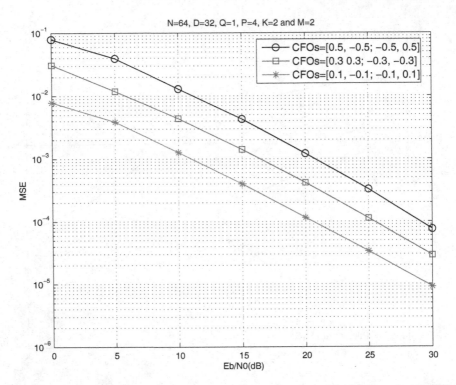

Fig. 5.4 Impact of different sets of CFOs on the MSE performance of the SOP based multi-CFO estimation method for multiuser CoMP OFDM systems, with $P = 4$ pilot blocks, $K = 2$ users and $M = 2$ BSs

This short pilot is carefully designed to achieve two main aims: (a) simultaneous low-complexity multi-CFO estimation at all BSs, and (b) permutation and quadrant ambiguities elimination for the ICA equalized signals. A cost function is formed for the design of these pilot symbols. By minimizing the cost function, OP and SOP are designed, respectively. OP is a set of pilots that are orthogonal to each other in the frequency and user domains. However, the minimum required length of the OP is large when the numbers of subcarriers and users are large. To allow a good trade-off between spectral efficiency, CFO estimation accuracy and performance of ICA ambiguity elimination, a number of pilots are designed to be orthogonal in the user domain, and semi-orthogonal in the frequency domain. They are referred to as SOP. The requirement on pilot length for the SOP is also discussed.

The precoding based CFO estimation method proposed in the previous chapter is only used for single CFO estimation, not for multi-CFO estimation. In order to achieve the purpose, the SOP based multi-CFO estimation method is proposed for CoMP OFDM systems. First, multiple CFOs are separated using the cross-correlation between the different pilots of users. A complex multi-dimensional search is divided into a number of one-dimensional searches at each BS. Second,

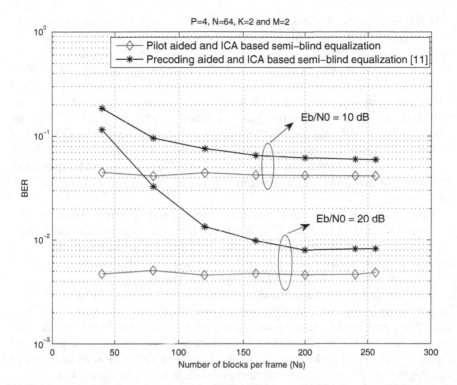

Fig. 5.5 Impact of the frame length on the BER performances of ICA based semi-blind equalization structures with SOP and precoding, and in the absence of CFO, with $K = 2$ users and $M = 2$ BSs

the proposed CFO estimation is to minimize the sum power of selected off-diagonal elements in the auto-correlation of the SOP. To allow a trade-off between complexity and performance, only a selected number of correlation coefficients are considered. Third, by taking the carrier frequency of one BS as a reference, a CFO correction method is proposed in which the CFO estimates only need to be sent to a reference BS for the calculation of how much carrier frequency should be adjusted at each user and each of the rest of BSs. As an additional receiver is not required to be installed at a BS to listen to each other for direct frequency synchronization, the system burden is low.

After CFO adjustments, ICA is employed for equalization at BSs on each subcarrier. The correlation between the equalized and original pilots is used to eliminate the ambiguity in the ICA equalized signals. By carefully designing the SOP, the cross-correlation between different pilots of users is minimized. This provides an optimal solution to the ambiguity elimination, by maximizing the real part of the cross-correlation between the equalized and original pilots. The proposed approach has lower complexity than the precoding based method, as the permutation and quadrant ambiguities can be eliminated simultaneously, rather than sequentially.

In the simulation, the proposed multi-CFO estimation approach is shown to provide better BER and MSE performances than Moose's method [3] and the CAZAC sequences based method [7]. With a low training overhead of 1.6%, the SOP based multi-CFO estimation and ICA equalization scheme can provide a BER performance close to the ideal case with perfect CSI and no CFO. The SOP aided semi-blind equalization scheme is shown to be more robust against channel variations than the precoding aided method in [11]. The proposed structure allows easy scalability and trade-off between performance, complexity and spectral efficiency. In particular, compared to the special case of OP, the SOP based scheme can achieve a 32-fold reduction in spectral overhead and a 2400-fold reduction in complexity of multi-CFO separation and estimation, while providing a close BER performance.

Although the SOP based CFO estimation method can be performed in the case with multiple CFOs, a feedback CFO correction is required between BSs and users. This results in an additional system overhead, which is not spectrally efficient. In the next chapter, a semi-blind joint ICI mitigation and equalization structure is proposed for CA based CoMP OFDMA systems. This proposed structure requires no overhead for CFO correction, and thus achieves higher spectral efficiency. Also, the previously proposed CFO estimation approaches are based on searching within a range of possible CFO values, resulting in higher complexity. In the next chapter, the semi-blind structure allows the CFO-induced ICI mitigation to be performed implicitly via semi-blind equalization. As a large number of searches are not required, the proposed system in the next chapter can have low complexity.

Appendix

Proof of Theorem 1. This appendix contains the proof that maximizing the cross-correlation $\rho_{\pi_k(n),k}(n)$ in Eq. (5.16) is a solution to the problem of the permutation and quadrant ambiguities.

By using the correlation in Eq. (5.16), the cross-correlation between the $\pi_k(n)$-th ICA separated pilot $\check{s}_{\pi_k(n)}(n, i)$ and the k-th user's pilot on subcarrier n over P blocks in the noiseless case can be written as

$$\frac{1}{P}\sum_{i=0}^{P-1}\check{s}_{\pi_k(n)}(n, i)s_k^*(n, i) = \begin{cases} e^{-j\theta} & \pi_k(n) = k \\ 0 & \pi_k(n) \neq k \end{cases}. \tag{5.18}$$

Substituting Eqs. (5.18) to (5.16) yields to

$$\rho_{\pi_k(n),k}(n) = \begin{cases} e^{j(\tilde{\theta}_{\pi_k(n)} - \theta)} & \pi_k(n) = k \\ 0 & \pi_k(n) \neq k \end{cases}. \tag{5.19}$$

Since the phase rotation $\theta \in \{0, \frac{\pi}{2}, \pi, \frac{3\pi}{2}\}$ and the trial rotation $\tilde{\theta}_{\pi_k(n)} \in \{0, \frac{\pi}{2}, \pi, \frac{3\pi}{2}\}$, their substraction is equal to $(\tilde{\theta}_{\pi_k(n)} - \theta) \in \{0, \pm\frac{\pi}{2}, \pm\pi\}$. As a result, $e^{j(\tilde{\theta}_{\pi_k(n)} - \theta)} \in \{1, -1, j, -j\}$. Only when the phase rotation is found as $\tilde{\theta}_{\pi_k(n)} = \theta$, the cross-correlation $\rho_{\pi_k(n),k}(n)$ becomes one. This also provides a solution to the permutation ambiguity problem, as the phase rotation can only be found on the correct substream as $\pi_k(n) = k$. The real part of $\rho_{\pi_k(n),k}(n)$ is maximized, while the imaginary part of $\rho_{\pi_k(n),k}(n)$ reduces to zero. Therefore, in the presence of noise, the real part of cross-correlation $\rho_{\pi_k(n),k}(n)$ can be maximized to eliminate the permutation and phase ambiguities simultaneously.

References

1. 3gpp technical report 36.814 version 9.0.0, further advancements for e-utra physical layer aspects, Mar. 2010.
2. M. Morelli and U. Mengali. Carrier-frequency estimation for transmissions over selective channels. *IEEE Transactions on Communications*, 48(9):1580–1589, Sep. 2000.
3. P. H. Moose. A technique for orthogonal frequency division multiplexing frequency offset correction. *IEEE Transactions on Communications*, 42(10):2908–2914, Oct. 1994.
4. M. Movahhedian, Y. Ma, and R. Tafazolli. Blind CFO estimation for linearly precoded OFDMA uplink. *IEEE Transactions on Signal Processing*, 58(9):4698–4710, Sep. 2010.
5. Y. Tsai, H. Huang, Y. Chen, and K. Yang. Simultaneous multiple carrier frequency offsets estimation for coordinated multi-point transmission in OFDM systems. *IEEE Transaction on Wireless Communications*, 12(9):4558–4568, Sep. 2013.
6. Y. Tsai, H. Huang, Y. Chen, and K. Yang. Simultaneous carrier frequency offset estimation for multi-point transmission in OFDM systems. In *Proc. IEEE Global Telecommuniation Conference (Globecom)*, Huston, USA, Dec. 2011.
7. Y. Wu, J. W. M. Bergmans, and S. Attallah. Carrier frequency offset estimation for multiuser MIMO OFDM uplink using CAZAC sequences: performance and sequence optimization. *EURASIP Journal on Wireless Communication and Networking*, 570680-1/11, 2011.
8. J. Hoydis, M. Kobayashi, and M. Debbah. Optimal channel training in uplink network MIMO systems. *IEEE Transactions on Signal Processing*, 59(6):2824–2834, 2011.
9. J. Karhunen A. Hyvarinen and E. Oja. *Independent Component Analysis*. John Wiley & Sons, New York, USA, May 2002.
10. L. Sarperi, X. Zhu, and A. K. Nandi. Blind OFDM receivers based on independent component analysis for multiple-input multiple-output systems. *IEEE Transactions on Wireless Communications*, 6(11):4079–4089, Nov. 2007.
11. J. Gao, X. Zhu, and A. K. Nandi. Non-redundant precoding and PAPR reduction in MIMO OFDM systems with ICA based blind equalization. *IEEE Transactions on Wireless Communications*, 8(6):3038–3049, Jun. 2009.
12. L. Weng, E. K. S. Au, P. W. C. Chen, R. D. Murch, R. S. Cheng, W. H. Mow, and V. K. N. Lau. Effect of carrier frequency offset on channel estimation for SISO/MIMO-OFDM systems. *IEEE Transactions on Wireless Communications*, 6(5):1854–1863, May 2007.
13. Q. K. Trinh and P. Z. Fan. Construction of multilevel Hadamard matrices with small alphabet. *Electronic Letter*, 44(21):1250–1252, Oct. 2008.
14. A. C. Lossifides. Complex orthogonal coded binary transmission with amicable Hadamard matrices over rayleigh fading channels. In *Proc. IEEE Symposium on Computers and Communications*, pages 335–340, Kerkyra, Greece, Jun. 2011.

15. B. W. Zarikoff and J. K. Cavers. Coordinated multi-cell systems: carrier frequency offset estimation and correction. *IEEE Journal on Selected Areas in Communications*, 28(9): 1490–1501, Dec. 2010.
16. J. F. Cardoso. High-order contrasts for independent component analysis. *Neural Computation*, 11(1):157–192, Jan. 1999.
17. A. Goldsmith. *Wireless Communications*. Cambridge University Press, London, U.K., 2005.
18. A. Cichocki and S. Amari. *Adaptive Blind Signal and Image Processing*. John Wiley, Chichester, U.K., 2003.
19. Y. Jiang, M. K. Varanasi, and J. Lin. Performance analysis of ZF and MMSE equalizers for MIMO systems: an in-depth study of high SNR regime. *IEEE Transaction on Information Theory*, 57(4):2008–2026, Apr. 2011.

Chapter 6
Semi-Blind Joint ICI Mitigation and Equalization for CA Based CoMP OFDMA Systems

6.1 Introduction

Although CoMP transmission [1] can reduce the interference between cells by connecting geographically separated BSs for cooperative communication, it is not enough to meet one of targets for LTE-Advanced, expected to provide significant improvement in cell-edge performance [1]. CA [1], where multiple component carriers of smaller bandwidth are aggregated for the concurrent transmission, can support very high data rates. Thus, CA transmission can be effectively employed in CoMP systems to improve cell-edge throughput [2]. So far, both technologies have been adopted by LTE-Advanced standards [1]. However, there has been little work to combine them in the literature.

As discussed in Sect. 2.2, one kind of CFO is the typical mismatch in LOs between transmitter and receiver, and destroys the orthogonality between subcarriers. In CoMP OFDMA systems, there are multiple CFOs. Apart from the ICI, these CFOs result in additional MUI. The previously proposed CFO estimation approaches require a large number of searches to obtain an accurate estimate of CFO, leading to high complexity. In the literature, several multi-CFO estimation methods have been proposed so far. In [3], a multi-CFO estimation method was proposed using CAZAC sequences. In [4] and [5], a number of CAZAC sequences were applied to CoMP systems, based on the ML algorithm. However, the above approaches also require search based algorithms to find an accurate estimate of CFO, leading to high computational complexity in the case with multiple CFOs.

With respect to the frequency location, component carriers can be categorized into three cases: intra-band continuous CA, intra-band non-continuous CA and inter-band non-continuous CA. One of targets for LTE-Advanced is to provide improvement in cell-edge spectral efficiency [1]. To meet the target, the inter-band non-continuous CA is considered, since other available free frequency bands can be used for the concurrent transmission to improve throughput of cell-edge users

© The Author(s) 2015
Y. Jiang et al., *Semi-Blind Carrier Frequency Offset Estimation and Channel Equalization*, SpringerBriefs in Electrical and Computer Engineering, DOI 10.1007/978-3-319-24984-1_6

in CoMP systems. Apart from the multiple CFOs caused by multiple users and BSs, additional CFOs are generated by extra LOs that are allowed to be installed. The signals transmitted on different component bands could experience different synchronization errors [2]. So far, very few papers have focused on multi-CFO estimation in the CA based system. In [6], a block type pilot based synchronization algorithm was proposed to estimate the CFO-induced ICI components. This ICI suppression is made by inverting a matrix with large size. In [7], two Multiple Signal Classification (MUSIC) based algorithms were proposed to estimate multiple CFOs by a set of preambles. However, these approaches could not be applied directly in the CA based CoMP OFDMA system which have a large number of CFOs. Also, they do not consider joint ICI mitigation and equalization.

In this chapter, a semi-blind ICA based equalization structure is presented for CA based CoMP OFDMA systems with multiple CFOs-corrupted signals.

- First, each OFDM block is divided into a number of subblocks such that the CFO-induced ICI is mitigated implicitly via semi-blind equalization. Thus, additional feedback overhead is not required. This is different from traditional methods where multiple CFOs are estimated first in OFDMA systems, and then the estimate values are fed back for CFO adjustments.
- Second, ICA [8], one of HOS based BSS techniques, is employed to equalize the CFO-corrupted received signals in the frequency domain, by exploiting the statistic characteristics of the received signals.
- Third, an ambiguity elimination approach is proposed to eliminate the permutation and quadrant ambiguities in the ICA equalized signals simultaneously by a small number of pilots, exploring the cross-correlation between the original and equalized pilot symbols.
- Fourth, the interference between subblocks is introduced in the proposed structure, which can be reduced by the proposed Successive Interference Cancellation (SIC) method and the subblock allocation scheme.
- Simulation results show that with a low training overhead, the proposed semi-blind ICA based semi-blind joint equalization and ICI mitigation scheme can provide a BER performance close to the ideal case with perfect CSI and no CFO.

The system model is presented in Sect. 6.2. The ICA based semi-blind ICI mitigation and equalization structure is proposed in Sect. 6.3. Complexity analysis is described in Sect. 6.4, and simulation results are shown in Sect. 6.5. Section 6.6 provides a summary.

6.2   System Model

Throughout the chapter, a CA based CoMP OFDMA system is considered in the uplink with M separated BSs, K users and N_c component carriers, as the block diagram as illustrated in Fig. 6.1. It is assumed that each component carrier has a specific frequency band. It is also assumed that each user and each BS

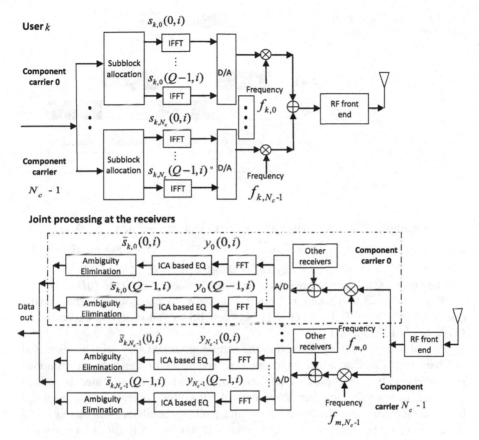

Fig. 6.1 Block diagram of the CA-CoMP OFDMA system with semi-blind ICA based receivers (EQ: equalization)

are equipped with a single antenna. This could be easily extended to multiple antennas. A number of BSs cooperatively process the symbols of unit streams simultaneously transmitted by cell-edge users, to improve system throughput and avoid the interference between cells. There are a total number of N_s OFDMA blocks, each consisting of N subcarriers. Each OFDMA block is partitioned into $Q = N/D$ subblocks, with D denoting the subblock length. The individual subblock at the transmitter is precoded by a D-point IDFT matrix $\mathbf{F}$, with entry (u, v) given by $\mathbf{F}(u, v) = 1/\sqrt{D}e^{-j2\pi uv/D}$, $(u, v = 0, 1, \ldots, D - 1)$.

Due to LOs mismatch between BSs and users, multiple CFOs exist in the CoMP OFDMA system. The application of CA in the CoMP OFDMA system may generate extra CFOs, because CA can allow additional LOs to be installed for the concurrent transmission, by using component carriers in different bands. Thus, there are a total number of MKN_c CFOs in the CA based CoMP OFDMA system. Define ϕ_{m,k,n_c} as the normalized CFO at the n_c-th component carrier ($n_c = 0, 1, \ldots, N_c - 1$) between the m-th BS ($m = 0, 1, \ldots, M - 1$) and the k-th user ($k = 0, 1, \ldots, K - 1$).

Fig. 6.2 Subblock allocation with $Q = 4$ subblocks and $K = 2$ users

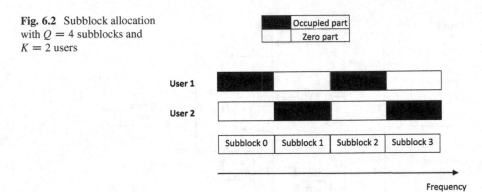

Correspondingly, the diagonal CFO matrix can be written as $\boldsymbol{\Phi}_{m,k,n_c} = \mathrm{diag}\{[1, w_{m,k,n_c}, \cdots, w_{m,k,n_c}^{N-1}]\}$, with $w_{m,k,n_c} = e^{\frac{j2\pi\phi_{m,k,n_c}}{N}}$. The CFO matrix for the q-th subblock is expressed as $\boldsymbol{\Phi}_{m,k,n_c}(q) = \mathrm{diag}\{[w_{m,k,n_c}^{qD}, w_{m,k,n_c}^{qD+1}, \cdots, w_{m,k,n_c}^{qD+D-1}]\}$, as a submatrix of $\boldsymbol{\Phi}_{m,k,n_c}$, with rows and columns obtained from qD to $(qD + D - 1)$ of $\boldsymbol{\Phi}_{m,k,n_c}$. The CFO-induced ICI matrix at the n_c-th component carrier in the q-th subblock ($q = 0, 1, \ldots, Q - 1$) between BS m and user k is expressed as $\mathbf{C}_{m,k,n_c}(q) = \mathbf{F}\boldsymbol{\Phi}_{m,k,n_c}(q)\mathbf{F}^H$.

The subband CAS is considered in the proposed OFDMA system, as discussed in Chap. 2. It is indicated that each subblock, consisting of some short continuous subcarriers, is only occupied by one user, and could not be allocated by others as shown in Fig. 6.2. However, this subblock allocation scheme causes the MUI between subblocks. In order to reduce the interference, the adjacent subblocks could not be occupied by the same user. That means if subblock q is allocated to user k, subblocks $(q - 1)$ and $(q + 1)$ could not be occupied by user k.

The channel is assumed to be quasi-static block fading, where the CSI remains constant for a frame duration of the N_s OFDMA blocks. The $L \times 1$ channel vector can be expressed as $[h_{m,k,n_c}(0), h_{m,k,n_c}(1), \ldots, h_{m,k,n_c}(L - 1)]^T$, with $h_{m,k,n_c}(l)$ denoting the l-th channel path ($l = 0, 1, \ldots, L - 1$) at the n_c-th component carrier between the m-th BS and the k-th user. It is assumed that each user uses the Zero Padding (ZP), instead of the CP, because ZP-OFDMA can explore the frequency diversity to combat the deep fading in some subcarriers [9], while the transmitted signals in the deep fading subcarrier may be lost in CP-OFDMA systems [10]. In ZP-OFDMA systems, a ZP of length L_{ZP} ($L_{ZP} \geqslant L - 1$) is appended to the time-domain signals during the guard interval before transmission, and removed at the receiver to avoid the ISI.

Let $s_{k,n_c}(n, q, i)$ denote the symbol on the n-th subcarrier ($n = 0, 1, \ldots, D - 1$) in the q-th subblock of the i-th OFDMA block ($i = 0, 1, \ldots, N_s - 1$) at the n_c-th component carrier, and transmitted by the k-th user. The transmitted data vector is defined as $\mathbf{s}_{k,n_c}(q, i) = [s_{k,n_c}(0, q, i), s_{k,n_c}(1, q, i), \ldots, s_{k,n_c}(D - 1, q, i)]^T$. Following [11], the received signals vector $\mathbf{y}_{m,n_c}(q, i)$ at the m-th BS combined with K users' signals in subblock q of OFDMA block i in the frequency domain is given by

$$\mathbf{y}_{m,n_c}(q,i) = \underbrace{\sum_{k=0}^{K-1} \mathbf{C}_{m,k,n_c}(q)\mathbf{H}_{m,k,n_c}^{(f)}\mathbf{s}_{k,n_c}(q,i)}_{\text{Desired signal + ICI}}$$

$$+ \underbrace{\sum_{k=0,q\neq0}^{K-1} \mathbf{C}_{m,k,n_c}(q)\check{\mathbf{H}}_{m,k,n_c}^{(f)}\mathbf{s}_{k,n_c}(q-1,i)}_{\text{MUI}} \qquad (6.1)$$

$$+ \underbrace{\mathbf{z}_{m,n_c}(q,i)}_{\text{noise}},$$

where $\mathbf{H}_{m,k,n_c}^{(f)} = \mathbf{F}\mathbf{H}_{m,k,n_c}^{(t)}\mathbf{F}^H$ is the channel frequency response matrix, with $\mathbf{H}_{m,k,n_c}^{(t)}$ denoting the $D \times D$ Toeplitz channel matrix, expressed as

$$\mathbf{H}_{m,k,n_c}^{(t)} = \begin{bmatrix} h_{m,k,n_c}(0) & 0 & \cdots & \cdots & 0 \\ \vdots & \ddots & 0 & \cdots & 0 \\ h_{m,k,n_c}(L-1) & \cdots & h_{m,k,n_c}(0) & & 0 \\ \vdots & \ddots & & \ddots & 0 \\ 0 & \cdots & h_{m,k,n_c}(L-1) & \cdots & h_{m,k,n_c}(0) \end{bmatrix}, \qquad (6.2)$$

$\check{\mathbf{H}}_{m,k,n_c}^{(f)} = \mathbf{F}\check{\mathbf{H}}_{m,k,n_c}^{(t)}\mathbf{F}^H$ is the interference matrix, with $\check{\mathbf{H}}_{m,k,n_c}^{(t)}$ being

$$\check{\mathbf{H}}_{m,k,n_c}^{(t)} = \begin{bmatrix} 0 \cdots h_{m,k,n_c}(L-1) & \cdots & \cdots & h_{m,k,n_c}(1) \\ 0 \cdots & 0 & h_{m,k,n_c}(L-1) \cdots & h_{m,k,n_c}(2) \\ \vdots & \cdots & \cdots & 0\ h_{m,k,n_c}(L-1) \\ 0 & \cdots & \cdots & 0 \\ \vdots & \vdots & \vdots & \vdots \\ 0 & \cdots & \cdots \cdots & 0 \end{bmatrix}, \qquad (6.3)$$

and $\mathbf{z}_{m,n_c}(q,i)$ is the AWGN vector whose entries are i.i.d. complex Gaussian random variables with a zero mean and a variance of N_0. The first term is the desired signals which suffer the ICI within each subblcok, the second term is the MUI from another user's $(q-1)$-th subblock, and the last term is the noise.

Remark 1. In the above system description, the CFO gives rise to the ICI, which is constrained in each subblock. The component on the n-th subcarrier suffers the interference from other $(D-1)$ frequency components. As the ICI term can be incorporated into the channel frequency response, the received signals are a linear mixture of the transmitted signals in each subblock in the frequency domain. Thus, ICA can be employed directly to the received signals in each subblock.

Remark 2. The q-th subblock is affected by the interference from another user's $(q-1)$-th subblock, except for the first subblock i.e., $q = 0$, which is free from the interference due to the use of the ZP. The subband CAS can reduce the interference to the neighbor's subblock of another user, while the interference between subblocks can be further reduced by the proposed SIC method, and the remaining is considered as noise which can be eliminated by the phase correction in the ICA process.

Remark 3. A number of multi-CFO estimation approaches have been proposed in the literature. However, it is very hard to compensate for multiple CFOs at the receiver [12, 13]. If not impossible, a feedback overhead is required, which results in low spectral efficiency. Also, the use of CA could lead to different CFOs on different component carriers, which makes multi-CFO estimation and compensation become much more complex.

6.3 ICA Based Equalization and ICI Mitigation

A semi-blind structure is proposed for CA based CoMP OFDMA systems, where the ICA based equalization and ICI mitigation can be performed jointly. As the CFO-induced ICI allows the received signals to become a linear mixture of the transmitted signals, the ICA based equalization can be employed to recover the CFO-corrupted received signals in each subblock. The subband CAS introduces the interference to the next subblock, which can be reduced by the proposed SIC method. Generally, equalization, ICI mitigation and SIC are repeated a number of times until the last subblock. This is different from other approaches which only consider CFO estimation, CFO compensation and equalization, separately rather than jointly.

6.3.1 ICA Based Equalization

In CoMP systems, the separated BSs are connected together to cooperatively process the received signals. Thus, the received signals from all M BSs are combined together as $\mathbf{y}_{n_c}(q,i) = \sum_{m=0}^{M-1} \mathbf{y}_{m,n_c}(q,i)$. This does not affect the structure of the received signals which are a linear mixture of the transmitted signals in each subblock. Therefore, ICA [8], an efficient HOS based BSS technique, can be employed on each subblock to obtain the estimate of the source signals, by maximizing the non-Gaussianity of the received signals $\mathbf{y}_{n_c}(q,i)$. JADE [14], based on the joint diagonalization of cumulant matrices of the received signals, requires shorter data sequences than other ICA numerical algorithms, and is used in this chapter. First, the received signals are whitened. Then, the JADE algorithm is used to obtain estimates of the source symbols by diagonalizing the cumulant matrices of the received signals in each subblock. After using ICA on the received signals, the equalized signals vector $\tilde{\mathbf{s}}_{k,n_c}(q,i)$ is expressed as

$$\tilde{s}_{k,n_c}(q,i) = \mathbf{G}_{k,n_c}(q)s_{k,n_c}(q,i), \tag{6.4}$$

where $\mathbf{G}_{k,n_c}(q)$ is the ambiguity matrix in the q-th subblock, given by

$$\mathbf{G}_{k,n_c}(q) = \mathbf{L}_{k,n_c}(q)\mathbf{P}_{k,n_c}(q)\mathbf{B}_{k,n_c}(q), \tag{6.5}$$

where $\mathbf{L}_{k,n_c}(q)$ denotes the phase deviation matrix, $\mathbf{P}_{k,n_c}(q)$ is the permutation ambiguity matrix, and $\mathbf{B}_{k,n_c}(q)$ is the quadrant ambiguity matrix.

6.3.2 Phase Correction

There are some possible phase deviations in the ICA equalized signal $\tilde{s}_{k,n_c}(q,i)$, which can be corrected by a de-rotation process. Let $\boldsymbol{\pi} = [\pi_0, \pi_1, \ldots, \pi_{D-1}]^T$ denote the permuted order. The ICA equalized signals vector can be written as $\tilde{\mathbf{s}}_{k,n_c}(q,i) = [\tilde{s}_{k,n_c}(\pi_0,q,i), \tilde{s}_{k,n_c}(\pi_1,q,i), \ldots, \tilde{s}_{k,n_c}(\pi_{D-1},q,i)]^T$. After phase correction, the k-th user's signals vector $\check{\mathbf{s}}_{k,n_c}(q,i) = [\check{s}_{k,n_c}(\pi_0,q,i), \check{s}_{k,n_c}(\pi_1,q,i), \ldots, \check{s}_{k,n_c}(\pi_{D-1},q,i)]^T$ in the q-th subblock is given by

$$\check{\mathbf{s}}_{k,n_c}(q,i) = \mathbf{L}_{k,n_c}^{-1}(q)\tilde{\mathbf{s}}_{k,n_c}(q,i) \tag{6.6}$$

where $\mathbf{L}_{k,n_c}^{-1}(q) = \mathrm{diag}\{[\frac{\alpha_{k,n_c}(\pi_0,q)}{|\alpha_{k,n_c}(\pi_0,q)|}, \ldots, \frac{\alpha_{k,n_c}(\pi_{D-1},q)}{|\alpha_{k,n_c}(\pi_{D-1},q)|}]\}$, with $\alpha_{k,n_c}(\pi_n,q)$ denoting the de-rotation factor $(\pi_n = \pi_0, \pi_1, \ldots, \pi_{D-1})$, obtained from the statistical characteristics of $\tilde{s}_{k,n_c}(\pi_n,q,i)$ with the QPSK modulation, given by

$$\alpha_{k,n_c}(\pi_n,q) = \{(1/N_s)\sum_{i=0}^{N_s-1}[\tilde{s}_{k,n_c}(\pi_n,q,i)]^4\}^{-\frac{1}{4}}e^{j\frac{\pi}{4}}. \tag{6.7}$$

6.3.3 Permutation and Quadrant Ambiguities Elimination

However, this phase correction brings a phase rotation in $\check{s}_{k,n_c}(q,i)$, which can be eliminated together with the left permutation ambiguity by a small number of pilot symbols of length P, exploring the cross-correlation between the equalized and original pilot symbols. The cross-correlation $\rho_{\pi_n,n}(q)$ between the π_n-th ICA separated pilot symbol $\check{s}_{\mathrm{pil},k,n_c}(\pi_n,q,i)$ and the original pilot symbol $s_{\mathrm{pil},k,n_c}(n,q,i)$ over P blocks in the q-th subblock is defined as

$$\rho_{\pi_n,n}(q) = \frac{1}{P}\sum_{i=0}^{P-1}\{[\check{s}_{\mathrm{pil},k,n_c}(\pi_n,q,i)e^{j\theta_{\pi_n}}]s_{\mathrm{pil},k,n_c}^*(n,q,i)\}, \tag{6.8}$$

with $\theta_{\pi_n} = \{\frac{\pi}{2}, \pi, \frac{3\pi}{2}, 2\pi\}$ denoting the possible phase rotation. The real part of the cross-correlation will increase as the symbols are decoded correctly, and $\rho_{\pi_n,n}(q)$ becomes real only when there is no error. Therefore, the real part of $\rho_{\pi_n,n}(q)$ is used to find the correct order of π_n^{cor} and the phase rotation of $\theta_{\pi_n^{cor}}$ simultaneously for the π_n^{cor}-th separated substream, by maximizing the following cost function

$$[\pi_n^{cor}, \theta_{\pi_n^{cor}}] = \arg \max_{\pi_n, \theta_{\pi_n}} \Re e \left\{ \frac{1}{P} \sum_{i=0}^{P-1} \{[\breve{s}_{\text{pil},k,n_c}(\pi_n, q, i)\right.$$

$$\left. \cdot e^{j\theta_{\pi_n}}]s_{\text{pil},k,n_c}^*(n, q, i)\} \right\}. \tag{6.9}$$

π_n^{cor} is considered as the highest possibility of being the correct order, arranged into the $D \times D$ permutation ambiguity elimination matrix $\mathbf{P}^{-1}(q)$, with the π_n^{cor}-th row expressed as

$$[\mathbf{P}^{-1}(q)](\pi_n^{cor}, :) = [\mathbf{0}_{1 \times \pi_n^{cor}} \ 1 \ \mathbf{0}_{1 \times (D-\pi_n^{cor}-1)}], \tag{6.10}$$

and $\theta_{\pi_n^{cor}}$ is considered as the highest possibility of being the correct phase rotation, arranged in a diagonal matrix as

$$\mathbf{B}^{-1}(q) = \text{diag}\{[e^{j\theta_{\pi_0^{cor}}}, e^{j\theta_{\pi_1^{cor}}}, \dots, e^{j\theta_{\pi_{D-1}^{cor}}}]\}. \tag{6.11}$$

Then, the estimate of the source data $\hat{s}_{k,n_c}(q, i)$ can be given by

$$\hat{s}_{k,n_c}(q, i) = \mathbf{B}^{-1}(q)\mathbf{P}^{-1}(q)\breve{s}_{k,n_c}(q, i). \tag{6.12}$$

6.3.4   Successive Interference Cancellation

The presented structure introduces the interference between subblocks. Part of interference can be reduced by the subblock allocation scheme, while other parts can be cancelled by the proposed SIC method. The leftover interference is considered as noise, which can be mitigated by the phase correction during the ICA based equalization. The diagram of the proposed SIC method is shown in Fig. 6.3.

Because of the ZP, the first interference-free subblock is directly passed to the ICA based equalization. Then, SIC is employed to reduce the interference passed to the second subblock. Again, ICA is performed in the second subblock. This process is repeated a number of times until the last subblock.

Let $\tilde{\mathbf{H}}_{m,k,n_c}^{(t)}(q) = \mathbf{\Phi}_{m,k,n_c}(q)\mathbf{H}_{m,k,n_c}^{(t)}$ and $\tilde{\mathbf{H}}_{m,k,n_c}^{(f)}(q) = \mathbf{C}_{m,k,n_c}(q)\mathbf{H}_{m,k,n_c}^{(f)}$. After performing ICA and ambiguity elimination on the received signals, the estimate of the source data $\hat{s}_{k,n_c}(q, i)$ in the q-th subblock at the m-th BS can be used for the LS based channel estimation $\hat{\tilde{\mathbf{H}}}_{m,k,n_c}^{(f)}$ associated with the CFO-induced ICI, written as

Fig. 6.3 SIC diagram for CA
based CoMP OFDMA
systems

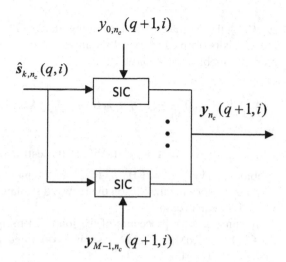

$$\hat{\tilde{\mathbf{\Pi}}}_{m,k,n_c}^{(f)}(q) = \mathbf{y}_{m,n_c}(q,i)[\hat{\mathbf{s}}_{k,n_c}(q,i)]^+, \tag{6.13}$$

where $(\cdot)^+$ denotes the pseudo-inverse. The estimate of the Toeplitz channel matrix
with the effect of CFO can be obtained by passing Eq. (6.13) to IDFT, given by

$$\hat{\tilde{\mathbf{H}}}_{m,k,n_c}^{(t)}(q) = \mathbf{F}^H \hat{\tilde{\mathbf{H}}}_{m,k,n_c}^{(f)}(q). \tag{6.14}$$

Toeplitz matrix is a special structure, where each row is equal to the previous one
cyclicly shifted by one element. As the CFO matrix is diagonal, the location of
zero-valued elements in $\mathbf{H}_{m,k,n_c}^{(t)}$ maintains in $\tilde{\mathbf{H}}_{m,k,n_c}^{(t)}(q)$. However, each row of the
Toeplitz channel matrix is cumulatively rotated by a phase of w_{m,k,n_c}, caused by the
CFO between BS m and user k at component carrier n_c. Therefore, the adjacent rows
of $\hat{\tilde{\mathbf{H}}}_{m,k,n_c}^{(t)}(q)$ can be explored to find the phase difference. By using all adjacent rows
of $\hat{\tilde{\mathbf{H}}}_{m,k,n_c}^{(t)}(q)$, the average estimate of the phase diffidence $\hat{w}_{m,k,n_c}$ is given by

$$\hat{w}_{m,k,n_c} = \frac{1}{D-1} \sum_{u=0}^{D-2} \frac{[\hat{\tilde{H}}_{m,k,n_c}^{(t)}(u+1,u+1,q)]}{[\hat{\tilde{H}}_{m,k,n_c}^{(t)}(u,u,q)]}, \tag{6.15}$$

where $\hat{\tilde{H}}_{m,k,n_c}^{(t)}(u,v,q)$ is the (u,v) entry of $\hat{\tilde{\mathbf{H}}}_{m,k,n_c}^{(t)}(q)$ in the q-th subblock. Correspondingly, the CFO matrix in the q-th subblock can be estimated as $\hat{\mathbf{\Phi}}_{m,k,n_c}(q) =$
$\mathrm{diag}\{[\hat{w}_{m,k,n_c}^{qD}, \hat{w}_{m,k,n_c}^{qD+1}, \cdots, \hat{w}_{m,k,n_c}^{qD+D-1}]\}$. The time-domain channel estimate can be
obtained as $\hat{\mathbf{H}}_{m,k,n_c}^{(t)} = \hat{\mathbf{\Phi}}_{m,k,n_c}^{-1}(q)\hat{\tilde{\mathbf{H}}}_{m,k,n_c}^{(t)}(q)$. In each column of $\hat{\mathbf{H}}_{m,k,n_c}^{(t)}$, there are
$\hat{L}$ non-zero elements while the rest $(D - \hat{L})$ elements are close to zero. Thus, the
first $\hat{L}$ elements in the first column of $\hat{\mathbf{H}}_{m,k,n_c}^{(t)}$ are selected as channel paths. By using

Eq. (6.3), the estimate of the interfering matrix $\hat{\overset{\circ}{\mathbf{H}}}{}^{(t)}_{m,k,n_c}$ is constructed. Next, the SIC method is performed to cancel the interference in $\mathbf{y}_{m,n_c}(q+1,i)$ in the $(q+1)$-th received subblock at the m-th BS as

$$\mathbf{y}_{m,n_c}(q+1,i) = \mathbf{y}_{m,n_c}(q+1,i) - \sum_{k=0,q\neq0}^{K-1} \hat{\mathbf{C}}_{m,k,n_c}(q)\hat{\overset{\circ}{\mathbf{H}}}{}^{(f)}_{m,k,n_c}\hat{\mathbf{s}}_{k,n_c}(q,i), \qquad (6.16)$$

where $\hat{\mathbf{C}}_{m,k,n_c}(q) = \mathbf{F}\hat{\boldsymbol{\Phi}}_{m,k,n_c}(q)\mathbf{F}^H$ is the estimate of the ICI matrix in the q-th subblock, and $\hat{\overset{\circ}{\mathbf{H}}}{}^{(f)}_{m,k,n_c} = \mathbf{F}\hat{\overset{\circ}{\mathbf{H}}}{}^{(t)}_{m,k,n_c}\mathbf{F}^H$ is the frequency-domain interfering matrix.

After performing the SIC on the received signals in the $(q+1)$-th subblock, ICA is used for equalization.

In summary, the procedure of the joint ICI mitigation and equalization scheme for CA based CoMP OFDMA systems is operated as follows:
For $q = 0 : Q - 1$

1. ICA is used for equalization by using Eq. (6.4).
2. The ambiguity elimination in the ICA equalized signals is performed by using Eqs. (6.6), (6.7), (6.8), (6.9), (6.10), (6.11), and (6.12).
3. The proposed SIC method is employed to cancel the MUI by using Eqs. (6.13), (6.14), (6.15), and (6.16).

End.

6.4   Complexity Analysis

In this section, the complexity of the proposed semi-blind ICA based joint equalization and ICI mitigation scheme is investigated for CA based CoMP OFDMA systems, in terms of the number of complex multiplications, as shown in Table 6.1. In the ICA based equalization, the main complexity is caused by the JADE method. Compared to the system with no CFO in [15], the proposed ICA based equalization structure has higher complexity, because a larger number of signals are put through the ICA process each time. Since the complexity of the proposed ICA based equalization grows with the length of subblock, D should be small. However, in order to reduce the interference between subblocks, the length of the subblock should be long. Therefore, there is a trade-off between interference reduction and complexity, in terms of the selection of the length of subblock.

In terms of ambiguity elimination, the proposed method requires a $\mathscr{O}(4PNN_c)$, while the precoding based approach in [15] requires a $\mathscr{O}(NN_c(N_s(D-1)!+1))$. Compared to the precoding based method, the proposed algorithm requires approximately $\frac{N_s(D-1)!+1}{4P}$ times less complexity, which depends mostly on the selection of the subblock length D as well as the frame length N_s. With $N_s = 256$, $D = 8$ and $P = 8$, this is equal to approximately 40,320 times less. One reason is

Table 6.1 Analytical computational complexity (EQ: equalization, est.: estimation)

Item		Order of complexity
IDFT		$N_s N_c N \log_2 D$
DFT		$N_s N_c N \log_2 D$
ICA EQ	ICA(JADE) [15]	$ND(1 + D^2 N_s + D^3)N_c$
	Phase correction	$NN_s N_c$
	Ambiguity elimination	$4PNN_c$
SIC	Channel est.	$N_s MNN_c D$
	IDFT	$MNN_c \log_2 D$
	Interference cancellation	$2N_s MNDN_c$

because the proposed ambiguity elimination only requires a small number of pilot symbols, rather than all the received signals as in [15]. Another reason is that the permutation and quadrant ambiguities problem can be resolved simultaneously by the proposed method, rather than sequentially by the precoding based approach. Also, the precoding based approach requires a process of encoding and decoding, which is not required by the proposed method.

The ICA based equalization and ICI mitigation are allowed to be performed jointly for CA based CoMP OFDMA systems, where the ICI is mitigated implicitly via the semi-blind equalization. This is different from the CFO estimation methods in [16] and [3], where a large number of searches and iterations are required to achieve an accurate CFO estimate. Therefore, the proposed structure can provide a lower complexity for the case with a large number of CFOs.

6.5 Simulation Results

In this section, simulation results are used to demonstrate the proposed ICA based joint ICI mitigation and equalization structure for CA based CoMP OFDMA systems, with $M = 2$ cooperative BSs and $K = 2$ users. There are a number of $N_c = 2$ non-continues component carriers, operating at 2.4 and 5.1 GHz, respectively. A frame consists of a total number of $N_s = 256$ OFDMA blocks with the QPSK modulation. Each OFDMA block has $N = 64$ subcarriers, divided into $Q = 8$ subblocks of a length $D = 8$. Clarke's block fading channel model [17] is employed, where the channel remains constant during a frame. The RMS delay spread is $T_{RMS} = 1.2$ normalized to the sampling time. The number of pilot symbols is $P = 8$, resulting in a low training overhead of 3.1%. The pilots are obtained from the well-studied Hadamard matrix where any two different rows are orthogonal to each other [18, 19]. The CFO is generated randomly in the range from -0.5 to 0.5.

Figure 6.4 demonstrates the BER performance of the semi-blind ICA based equalization and ICI mitigation for CA based CoMP OFDMA systems, compared to the performance of the ZF based equalization method with perfect CSI and no CFO. The CAZAC sequences based CFO estimation method [3] and the LS based channel

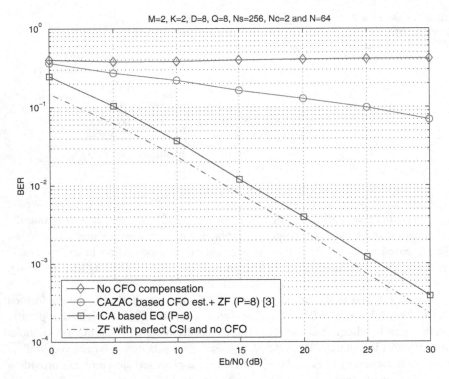

Fig. 6.4 BER performance of the ICA based joint equalization and ICI mitigation structure for CA based CoMP OFDMA systems with multiple CFOs (EQ: equalization)

estimation approach are also shown for comparison. The small performance gap between the proposed semi-blind structure and the ZF based equalization method with perfect CSI and no CFO is due to the leftover interference in the received signals. Without compensation, the ZF based equalization method has an error floor at high BER across the whole range of E_b/N_0s. A number of $P/2 = 4$ CAZAC sequences are used for multi-CFO estimation, while a number of $P/2 = 4$ training symbols are used for the LS based channel estimation plus the ZF based equalization. With the same number of pilot symbols, the proposed ICA based joint ICI mitigation and equalization structure provides a better BER performance than the CAZAC based method in [3]. Therefore, the proposed structure is shown to have a high performance.

Figure 6.5 demonstrates the impact of different CFOs sets on the BER performance for the proposed semi-blind ICA based joint equalization and ICI mitigation structure. Two extreme cases are considered for CFO sets $\phi_{m,k,n_c} = \pm 0.5$ and ± 0.1. These two CFO sets provide almost the same performance, which are close to the ZF based equalization method with perfect CSI and no CFO. Therefore, the proposed semi-blind ICA based joint equalization and ICI mitigation structure is shown to be robust against CFO variations.

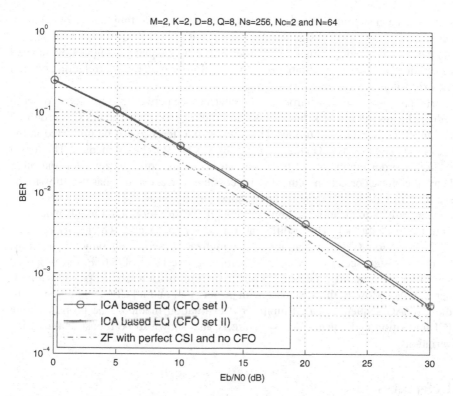

Fig. 6.5 Impact of different CFOs sets on BER performance of the ICA based equalization structure for CA based CoMP OFDMA systems (EQ: equalization, CFO set I: (component carrier 1: [0.1 −0.1; −0.1 0.1], component carrier 2: [0.1 0.1; −0.1 −0.1]), CFO set II: (component carrier 1: [0.5 −0.5; −0.5 0.5], component carrier 2: [0.5 0.5; −0.5 −0.5]))

6.6 Summary

In this chapter, a semi-blind ICA based joint ICI mitigation and equalization structure is proposed for CA based CoMP OFDMA systems, where the CFO-induced ICI is mitigated implicitly via semi-blind equalization. The application of CA in the CoMP OFDMA system may generate extra CFOs, because CA can allow additional LOs to be installed for the concurrent transmission, by using component carriers at different bands. The previously proposed CFO estimation approaches require a large number of searches to obtain an accurate estimate of CFO, which results in high complexity for the case with large number of CFOs. Traditionally, multiple CFOs are estimated first, and then the estimate values are fed back for CFO adjustments. The proposed structure requires no additional feedback overhead for CFO correction, and thus provides a low overhead.

Each OFDMA block is divided into a number of subblocks. In each subblock, the received signals become a linear mixture of the transmitted signals with the CFO-induced ICI. ICA can be employed to equalize the CFO-corrupted received signals in each subblock in the frequency domain, by exploiting the statistical characteristics of the received signals. However, the ICA based equalization introduces the permutation and quadrant ambiguities in the ICA equalized signals, which can be eliminated by a small number of pilots, exploring the cross-correlation between the original and equalized pilot symbols.

The proposed structure introduces the interference between subblocks. In order to reduce the interference, a subblock allocation scheme is used. Then, SIC is proposed to cancel the interference further. The leftover interference can be mitigated in the ICA based equalization. On the one hand, ICA is an HOS based method. The fourth or higher-order cumulants of the Gaussian noise are equal to zero. On the other hand, the phase correction in the ICA process can correct some phase shifting in the received signals caused by noise and leftover interference.

Generally, equalization, ICI mitigation and SIC are repeated a number of times until the last subblock. This is different from other approaches which only consider CFO estimation, CFO compensation and equalization, separately rather than jointly.

In the tested scenarios, with a low training overhead of 3.6 %, the proposed semi-blind ICA based joint equalization and ICI mitigation scheme provides a BER performance close to the ideal case with perfect CSI and no CFO, and outperforms the CAZAC sequences based multi-CFO estimation method. The proposed joint ICI mitigation and equalization structure is also shown to be robust against CFO variations.

References

1. 3gpp technical report 36.814 version 9.0.0, further advancements for e-utra physical layer aspects, Mar. 2010.
2. R. Ratasuk, D. Tolli, and A. Ghosh. Carrier aggregation in LTE Advanced. In *Proc. IEEE Vehicular Technology Conference*, pages 1–5, Taipei, Taiwan, May 2010.
3. Y. Wu, J. W. M. Bergmans, and S. Attallah. Carrier frequency offset estimation for multiuser MIMO OFDM uplink using CAZAC sequences: performance and sequence optimization. *EURASIP Journal on Wireless Communication and Networking*, 570680-1/11, 2011.
4. Y. Tsai, H. Huang, Y. Chen, and K. Yang. Simultaneous multiple carrier frequency offsets estimation for coordinated multi-point transmission in OFDM systems. *IEEE Transaction on Wireless Communications*, 12(9):4558–4568, Sep. 2013.
5. Y. Tsai, H. Huang, Y. Chen, and K. Yang. Simultaneous carrier frequency offset estimation for multi-point transmission in OFDM systems. In *Proc. IEEE Global Telecommuniation Conference (Globecom)*, Huston, USA, Dec. 2011.
6. X. Zhang, H. G. Ryu, and J. U. Kim. Suppression of synchronization errors in OFDM based carrier aggregation system. In *Proc. IEEE Asia-Pacific Conference*, pages 106–111, Auckland, New Zealand, Nov. 2010.
7. C. Y. Chang, Y. T. Sun, and M. L. Ku. MUSIC-based multiple CFOs estimation methods for CA-OFDM systems. In *Proc. IEEE International Conference on ITS Telecommunications*, pages 702–707, St. Petersburg, Russia, Aug. 2011.
8. J. Karhunen A. Hyvarinen and E. Oja. *Independent Component Analysis*. John Wiley & Sons, New York, USA, May 2002.
9. B. Muquet, Z. Wang, G. B. Giannakis, M. de Courville, and P. Duhamel. Cyclic prefixing of zero padding for wireless multicarrier transmissions? *IEEE Transactions on Communications*, 50(12):2136–2146, Dec. 2001.

10. J. D. Gibson. *The Communications Handbook*. CRC Press, London, U.K., 2002.
11. L. Weng, E. K. S. Au, P. W. C. Chen, R. D. Murch, R. S. Cheng, W. H. Mow, and V. K. N. Lau. Effect of carrier frequency offset on channel estimation for SISO/MIMO-OFDM systems. *IEEE Transactions on Wireless Communications*, 6(5):1854–1863, May 2007.
12. M. Morelli, C. C. J. Kuo, and M. O. Pun. Synchronization techniques for orthogonal frequency division multiple access (OFDMA): a tutorial review. *Proceedings of the IEEE*, 95(7): 1394–1427, Jul. 2007.
13. M. Movahhedian, Y. Ma, and R. Tafazolli. Blind CFO estimation for linearly precoded OFDMA uplink. *IEEE Transactions on Signal Processing*, 58(9):4698–4710, Sep. 2010.
14. J. F. Cardoso. High-order contrasts for independent component analysis. *Neural Computation*, 11(1):157–192, Jan. 1999.
15. J. Gao, X. Zhu, and A. K. Nandi. Non-redundant precoding and PAPR reduction in MIMO OFDM systems with ICA based blind equalization. *IEEE Transactions on Wireless Communications*, 8(6):3038–3049, Jun. 2009.
16. M. Morelli and U. Mengali. Carrier-frequency estimation for transmissions over selective channels. *IEEE Transactions on Communications*, 48(9):1580–1589, Sep. 2000.
17. A. Goldsmith. *Wireless Communications*. Cambridge University Press, London, U.K., 2005.
18. Q. K. Trinh and P. Z. Fan. Construction of multilevel Hadamard matrices with small alphabet. *Electronic Letter*, 44(21):1250–1252, Oct. 2008.
19. A. C. Lossifides. Complex orthogonal coded binary transmission with amicable Hadamard matrices over rayleigh fading channels. In *Proc. IEEE Symposium on Computers and Communications*, pages 335–340, Kerkyra, Greece, Jun. 2011.

Chapter 7
Conclusions

Several semi-blind CFO estimation and equalization schemes are proposed for a number of wireless communication systems. The statistical tool of ICA is used to perform semi-blind equalization in these systems which suffer the effect of CFO. These semi-blind systems have higher spectral efficiency than training based systems, since no or only few pilots are required.

In Chap. 4, a semi-blind precoding aided CFO estimation method and an ICA based equalization structure are proposed for single-user MIMO OFDM systems. A number of reference data sequences are superimposed into the source data via a linear precoding process, with no addition to the total transmit power consumed and no real-time spectral overhead introduced. The reference data sequences are carefully designed offline, which can kill two birds with one stone: CFO estimation and ambiguity elimination. A cost function is formed for the reference data sequences and the projection operator. By minimizing the cost function, the optimal reference data sequences and the projection operator can be designed, and obtained from a pool of orthogonal sequences in the Hadamard matrix. The precoding constant introduced in the precoding process is discussed, which provides a trade-off of power between the source data and the reference data. The precoding based CFO estimation is performed by minimizing the sum cross-correlations between the reference data sequences and the rest of orthogonal sequences in the pool. In general, the larger the number of selected sequences, the more accurate the CFO estimation. After CFO estimation and compensation, ICA is employed for equalization at the receiver on each subcarrier. However, there are permutation and phase ambiguities in the ICA equalized signals. According to the reference data sequences design, the cross-correlation between different reference data sequences of transmit antennas is minimized, to ensure that the different reference data sequences are orthogonal to each other. This provides an optimal solution to eliminate the permutation ambiguity by maximizing the cross-correlation between the ICA equalized signals and the reference signals. This correlation can be used again on each subcarrier for phase

© The Author(s) 2015
Y. Jiang et al., *Semi-Blind Carrier Frequency Offset Estimation and Channel Equalization*, SpringerBriefs in Electrical and Computer Engineering, DOI 10.1007/978-3-319-24984-1_7

ambiguity elimination. Simulation results show that the proposed semi-blind single-user MIMO OFDM system with precoding aided CFO estimation and ICA based equalization scheme, achieves a BER performance close to the ideal case with perfect CSI and no CFO.

In Chap. 5, a low-complexity semi-blind multiuser CoMP OFDM system is addressed, with a multi-CFO estimation method and an ICA based equalization scheme. A short pilot is carefully designed, killing two birds with one stone: simultaneous low-complexity multi-CFO estimation and ambiguity elimination for the ICA equalized signals. A cost function is formed for the design of these pilot symbols. By minimizing the cost function, OP and SOP are designed, respectively. OP is a set of pilots that are orthogonal to each other in the frequency and user domains. However, the minimum required length of the OP is large when the numbers of subcarriers and users are large. To allow a good trade-off between spectral efficiency, CFO estimation accuracy and performance of ICA ambiguity elimination, a number of pilots are designed to be orthogonal in the user domain, and semi-orthogonal in the frequency domain. They are referred to as SOP. First, multiple CFOs are separated by using the cross-correlation between different pilots of users. A complex multi-dimensional search for multiple CFOs is divided into a number of one-dimensional searches at each BS. The SOP based CFO estimation is to minimize the sum power of the selected off-diagonal elements in the auto-correlation of SOP. To allow a trade-off between complexity and performance, only a selected number of correlation coefficients are considered. By taking the carrier frequency of one BS as a reference, a CFO correction method is proposed that the CFO estimates only need to be sent to a reference BS for the calculation of how much carrier frequency should be adjusted at each user and each of the rest BSs. As additional receiver is not required to be installed at BS to listen to each other for direct frequency synchronization, the system burden is low. Second, the correlation between the equalized and original pilots is used to eliminate the ambiguity in the ICA equalized signals. By carefully designing the SOP, the cross-correlation between different pilots of users is minimized. This provides an optimal solution to the ambiguity elimination, by maximizing the real part of the cross-correlation between the equalized and original pilots. The SOP based approach has lower complexity than the precoding based method, as the permutation and quadrant ambiguities can be eliminated simultaneously, rather than sequentially. Simulation results show that, the proposed semi-blind multiuser CoMP OFDM system provides a BER performance close to the ideal case with perfect CSI and no CFO.

In Chap. 6, a semi-blind ICA based equalization and ICI mitigation structure is proposed for CA based CoMP OFDMA systems, where the CFO-induced ICI is mitigated implicitly via semi-blind equalization. ICA is employed to equalize the CFO-corrupted received signals in the frequency domain, by exploiting the statistical characteristics of the received signals. The ICA based equalization introduces ambiguity in the equalized signals, which can be eliminated by a small number of pilot symbols. Traditionally, multiple CFOs are estimated first, and then the estimate values are fed back to the transmit side for CFO adjustments. The proposed structure requires no additional feedback overhead for CFO correction, and thus provides a lower overhead. Each OFDMA block is divided into several subblocks. The proposed structure introduces the interference between subblocks. In order to reduce the interference, a subblock allocation scheme is first used. Then, SIC is proposed to cancel the interference further. The leftover interference can be mitigated in the ICA based equalization. Because ICA is an HOS based method, the fourth or higher-order cumulants of the Gaussian noise are equal to zero. Also, the phase correction in the ICA process can correct some phase shifting in the received signals caused by noise and left interference. Generally, equalization, ICI mitigation and SIC are repeated a number of times until the last subblock. This is different from other approaches which only consider CFO estimation, CFO compensation and equalization, separately rather than jointly. Simulation results show that the proposed semi-blind ICA based joint equalization and ICI mitigation scheme provides a BER performance close to the ideal case with perfect CSI and no CFO.

In summary, a number of ICA based semi-blind equalization and CFO estimation methods are proposed in this thesis for a number of wireless communication systems, to resolve the problems associated with CFO. The proposed semi-blind systems are a viable alternative to training based systems, as semi-blind systems can provide performances close to the ideal case with perfect CSI and no CFO at the receiver, while requiring little or no training overhead.

Printed in the United States
By Bookmasters

Printed in the United States
By Bookmasters